Physics
Made Simple

Ira M. Freeman

Revised by William J. Durden

Edited and prepared for publication by The Stonesong Press, Inc.

A Made Simple Book

Broadway Books

New York

A previous edition of this book was originally published in 1990 by Doubleday,
a division of Random House, Inc. It is here reprinted by arrangement with Doubleday.

Physics Made Simple. Copyright © 1990 by Doubleday,
a division of Random House, Inc.

Edited and prepared for publication by The Stonesong Press, Inc.
Managing Editor: Sheree Bykofsky
Editor: Lillian R. Rodberg
Editorial Consultant: James Madsen, Ph.D.
Design: Blackbirch Graphics, Inc.

Broadway Books titles may be purchased for business or promotional use or for special sales.
For information, please write to: Special Markets Department, Random House, Inc.,
1540 Broadway, New York, NY 10036.

MADE SIMPLE BOOKS and BROADWAY BOOKS are trademarks of Broadway Books,
a division of Random House, Inc.

Visit our website at www.broadwaybooks.com

First Broadway Books trade paperback edition published 2001.

The Library of Congress Cataloging-in-Publication Data has cataloged the Doubleday edition as:
Freeman, Ira Maximilian, 1905–.
    Physics made simple / Ira M. Freeman; revised by
    William J. Durden.—1st ed.
        p.   cm.
    Includes index.
    1. Physics.   I. Durden, William J.   II. Title.
QC23.F775   1990   89-1165
530—dc19   CIP
ISBN 0-385-24228-X

23 22 21 20 19 18 17 16 15

CONTENTS

PART ONE

MATTER AND ENERGY

CHAPTER 1

States of Matter

KEY TERMS FOR THIS CHAPTER

physics   element
matter   states of matter

What Is Physics?

This book introduces you to **physics,** the study of matter and energy and the relationships between them. These relationships are as old as the universe itself, and even though you may not realize it, the area of physics known as classical physics concerns itself with how things "work" in your everyday world. Every time you lift a barbell or a baby, drive a car or ride a bike, listen to your stereo, watch TV, dress in layers for a cold-weather jog, change the batteries in your flashlight, or make photocopies, you are demonstrating one or more of the principles of classical physics in action.

Classical physics, which you will be learning about in this book, is concerned with such topics as

o mechanics: motion and the forces that cause it
o acoustics: sound and the ways it is transmitted
o optics: light and its transmission
o thermodynamics: heat and energy

as well as electricity and magnetism. The principles you will learn here might be called "commonsense physics." They are as relevant to ordinary, everyday conditions as they were when first explained.

However, more recent scientific studies have shown that under extreme conditions, commonsense physics does not seem to apply. On a very small or a very vast scale, at extremely high or extremely low temperatures, or when moving at tremendous speed, matter and energy behave differently from what classical physics would lead one to expect.

Modern physics, which dates from about the close of the nineteenth century to the present, has developed such concepts as quanta, or particles; relativity of time, space, and motion; and solid state, to explain the relationships of matter and energy in extreme circumstances. New discoveries are continually being made. Although this book gives some glimpses into the world of modern physics, it is mainly devoted to helping you understand matter, energy, and concepts related to them as you encounter them every day.

States of Matter

We said at the outset that physics as a science concerns itself with matter and energy. Later, we will define and discuss energy; for now, what is matter? Briefly, you can think of **matter** as "anything that takes up space." Matter is everywhere: at home, where we work, in streets and forests, rivers and mountains. Although it seems as though matter takes a vast variety of forms, materials that we think of as being very different—wood, steel, glass, even water and air—are actually composites, or mixtures, of chemicals that are called **elements** because they cannot be broken down into further substances. Chemists have identified about a million compounds but only about one hundred elements.

Matter takes three basic forms called **states of matter:** solids, liquids, and gases. You can think of a *solid* as a form of matter that has both definite shape and volume—a bar of iron, a rock, an ice cube. A *liquid* has volume but no shape of its own; it takes the shape of whatever container it is in. Water is the most familiar liquid. A *gas* has neither definite shape nor definite volume. Steam is a gas, and so is the air all around you. Suppose you pump up a flat tire; the air you pump in (which is a gas) will fill the container (the tire) uniformly and assume both its volume and its shape.

Many common substances are mixtures of matter in several different states. For example, if you mix extremely fine sand, silt, or clay with water, the particles (which are solids) will practically never settle out. The water and the particles will form a cloudy liquid called a *colloidal suspension*—a stable mixture of a solid and a liquid. Ink is an example of a colloidal suspension; so are some common products such as antacids. Milk, on the other hand, is an *emulsion*—globules of one liquid (fat) suspended in another (water). Foam is formed by a gas (often air) suspended in a liquid.

Matter can change its state in response to certain conditions. Did you notice that in the example given above, water was listed in three different states: as water (a liquid), as ice cubes (a solid), and as steam

---

The three states of matter are solids, liquids, and gases. Matter may change from one state to another.

As a *general rule,* matter has permanence; it can take different forms but cannot stop existing.

(a gas). Water assumes these states in response to changes in temperature. We usually think of oxygen as a gas, but at about 300 degrees below zero Fahrenheit it turns into a bluish liquid. Iron, steel, and rock are solid as we normally see them, but all three will turn liquid if subjected to extremely high temperature. The lava that flows from an erupting volcano is superheated rock. In the sun and other stars, where temperatures reach thousands of degrees, iron exists as a gas.

The changes we have just described are called *physical changes;* the material changes its physical state but keeps its identifying characteristics. For example, ice, water, and steam are composed of the chemicals hydrogen and oxygen. Changes that occur in some materials affect their chemical characteristics and actually form a new substance. These kinds of changes are called *chemical changes.* In this book we will be talking about physical changes. Also, we will not be studying the properties, or characteristics, of matter that are known as *special properties.* That is what chemists are concerned with. We will be exploring the *general characteristics* common to all kinds of matter.

One general characteristic of matter is *permanence.* As a general rule, we can neither manufacture nor destroy matter; all we can do is change it from one form to another. That is, we can freeze water and turn it to ice, then turn the ice to steam by applying heat. Whatever its form, the water does not stop being a mixture of hydrogen and oxygen.

Another general characteristic of matter is that it takes up space. No two things can occupy the same space at the same time. A boat pushes aside water as it passes, and a chisel forces apart the fibers of a block of wood. Even air acts to keep other intruding material out (see Exploration 1.1).

Sometimes, two materials do seem to occupy the same space at the same time. For example, if you add instant coffee powder or soup mix to a cup of water, the liquid may not rise noticeably higher in the cup. That is because neither water nor any other substance is *continuous* matter. There are spaces between the water molecules into which other molecules such as that from the coffee powder can enter.

Exploration 1.1

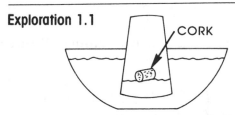

Fill a pan or a mixing bowl with water. Float a small cork on the water and push the open end of a water glass down over it. See how the cork changes position? The air inside the glass is pushing the water inside it down. The same principle applies to pumping air into the suit of a deep-sea diver (pressurizing the suit).

We will talk about other properties of matter, such as gravity, density, and inertia, in other chapters.

No two things can occupy the same space at the same time; if they seem to do so, it is because of the spaces between molecules that exist in all matter.

EVERYDAY PHYSICS: 1

"I can't get my eyes open," you say to yourself. "I need that cup of coffee—but there's hardly time before work." You run some water into the teakettle, turn on the burner, and rush off to finish dressing. The whistle of the teakettle brings you back to the kitchen at a run. You pour the boiling water over the instant coffee in your mug. "Darn! Now it's too hot to drink!" You grab a couple of ice cubes from the freezer and pop them into your coffee. They melt almost as soon as they hit the water and cool your coffee enough so you can drink it. From reading the preceding chapter, you know you've been seeing water assuming all three states of matter. You've also used principles of physics in (to name a few) drawing water from the tap, filling the kettle without splashing, heating the water, and, earlier, making the ice cubes. Physics is part of your everyday life in more ways than you probably guess. You'll know many of them when you complete working with this text.

Physics and Measurement

Physics is an exact science, meaning that the things physics deals with can be measured (quantified) precisely. From earliest times, people have found ways to specify quantities such as the distance to the next village; the interval of time between important events such as market days or festivals; or the amount of wheat, meat, cattle feed, or gold they bought and sold. To do this, they developed standard units of measure.

Some kinds of measurements are easy to make, especially if precision does not matter. How many hands high is the horse? How many suns (days) distant is the marketplace? But how do you measure fractions of a second? The intensity of light rays? Something so small you cannot see it, like a virus? Something immensely large or far away, like the sun? The study of physics involves measuring many things that are difficult to measure yet need to be precisely quantified.

Some things cannot be measured directly; instead, sophisticated techniques must be devised for measuring them indirectly. For example, physicists determine the size of an atom and its com-

---

Things that are very large or very small must often be measured by indirect means.

---

ponents by firing oppositely charged particles into the atom. The atom deflects them, and the size of the atom can then be estimated by the distance over which the particles are deflected. The sizes involved are so small that none of the particles can be seen. The physicist Albert A. Michelson first determined the speed of light by reflecting a flash of light off a rotating eight-sided mirror and a concave mirror. By measuring the distance the light traveled and the mirror's speed of rotation, he was able to calculate the speed of light. Today, we know that light is an electromagnetic wave and that its speed can be accurately measured by measuring the frequency of the waves.

Devising a means of measuring something and working out the appropriate unit sometimes takes years of research. Many units of measure used in physics and in everyday life are named in honor of the scientists who first devised them, including the hertz (Hz), which is used to measure electromagnetic frequencies; the joule, a unit of energy; and the angstrom, another electromagnetic measurement unit. In measuring something, we need to choose a unit that is proportional to what we are trying to measure; the inch would be inappropriate for measuring the distance from New York to California, and the mile would be inappropriate for measuring the living room for carpet. We also need a unit appropriate to the character-

istic we want to measure: a linear unit such as a foot or a mile for distance, a volume unit such as a gallon to measure liquids, a weight unit such as a pound or a ton for weight, and so on.

Certain units of measurement are called **fundamental units** because they can be combined to measure a wide variety of things. Examples of fundamental units include the foot, the mile, and the pound. **Derived units** are more complicated; they consist of combinations of fundamental units. In this chapter you will learn about measurements made in fundamental units: length, mass and weight, and time. You will also be learning about measurements made in derived units: area, volume, and density.

Measuring Length

Measuring length, or distance, requires a length unit, for example, the inch, the yard, the mile. You can select any unit you want to use, so long as you stick with it for all your calculations. In the English system of measure, which is also widely used in the United States, the *standard yard* is taken to be the distance between the end marks on a certain bronze bar kept in a vault at the Office of the Exchequer in London. (The yard is now defined in terms of the meter: 1 yard = 0.9144 meter.) In countries that use the English system, all goods sold by length, such as

---

Derived units of measurement consist of combinations of fundamental units; for example, measurements of area such as the square foot and the acre are multiples of length measurements.

In the metric system of measurement, all units are multiples of 10; that is, it is a decimal system.

fabric or rope, are measured by a stick, tape, or other device that has been marked off according to copies of the standard yard kept in the bureau of standards in each country.

One problem with the units and sizes of the different lengths used in the English system is that they don't seem to be related to one another in any clear way. We have to constantly remember that there are 12 inches in 1 foot, 3 feet in 1 yard, 5280 feet in 1 mile, and so on. Converting a measurement from one unit to another can be difficult; do you remember learning to multiply and divide feet and inches and the problems you had with carrying and remainders?

The **metric system** of measurement was worked out nearly two centuries ago to solve this difficulty. All units in the metric system are multiples of 10. To change one unit to another, all we have to do is to divide or multiply by 10. There are metric units for all the types of measurement we are talking about in this chapter, and the metric system is standard for all scientific work.

The fundamental length unit in the metric system is the standard **meter.** It was originally defined as the distance between the ends of a bar of platinum alloy kept at the International Bureau of Weights and Measures in France. Bureaus of standards in other countries keep precise copies of this bar. The meter (abbreviated as m) is a little longer than the yard—39.4 inches, to be exact—and consists of 100 centimeters (cm) and 1000 millimeters (mm). You can compare the centimeter with the inch in Figure 2.1. (The meter is now defined in terms of the speed of light in a vacuum.)

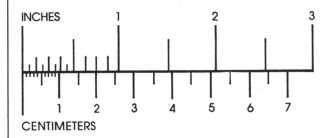

Figure 2.1

Table 2.1   Common Metric Units of Length

Metric Unit	Abbreviation	Number of Meters	English Equivalent
meter	**m**	**primary unit**	**39.4 inches**
kilometer	km	1000 m	0.621 mile
centimeter	cm	0.01 m	0.394 inch
millimeter	mm	0.001 m	0.039 inch
		0.100 cm	
2.54 cm			1.00 inch

Each division or multiple of a meter is named by putting a prefix to the word "meter"; a *centi*meter, from the Latin for "hundred," is $^1/_{100}$ (0.01) of a meter; a *milli*meter, from the Latin for "thousand," is $^1/_{1000}$ (0.100) of a meter and $^1/_{10}$ (0.010) of a centimeter; a *kilo*meter, from the Greek word for "thousand," is 1000 meters. The most common metric length units are shown in Table 2.1. Notice that metric measurements can be easily multiplied or divided by moving decimals.

Suppose the size of a rug had been given in kilometers as 0.0012 km. This is a small decimal, and it would be easier to visualize the rug in terms of a smaller unit—say, a meter. Since a kilometer equals 1000 meters, you would simply need to move the decimal point three places to the right (that is, multiply by 1000) to know that the rug is 1.2 meters long. If you wanted the rug's length in centimeters, you would move the decimals in 0.0012 km five places to the right (multiply by 100,000).

Now suppose that, instead, you wanted to convert a distance of 1.47 miles to inches. There are 12 inches to a foot and 5280 feet to a mile; you would have to multiply:

12 inches × 5280 feet × 1.47 miles

Your answer would be an unwieldy 93,139.2 inches (which you might round off to 93,100). Now, suppose you wanted to convert that to yards: You would have to divide by 36:

93,100 ÷ 36 = 2586

Exploration 2.1

You are on an aerobic exercise program, and on rainy days you do your jogging in the hallway of your apartment house. You measure the length of the hallway and find that it is 90 feet long. How many times will you need to run *up the hallway and back* to equal 1 mile?

Now, convert the length of the hall to meters. How many times must you run *up the hallway and back* to run 1 kilometer?

Which calculation was easier?

Measuring Area and Volume

Remember that to measure an attribute you need a unit of measure expressed in that same attribute—a length unit for length, for example. To measure **area**, a surface defined by given dimensions, you need an arbitrary unit that is itself an area. The formula for measuring the area of a square or rectangle is:

$$A \text{ (area)} = L \text{ (length)} \times W \text{ (width)}$$

To find a measurement that is itself an area, it is simplest to use a square with sides that equal a length unit—an inch, a yard, a meter, or whatever. In a square, length and width are the same. Multiplying a number by itself is to *square* the number. So, for area measurement, we have square inches, square feet, square meters, and so forth. A shorthand way for indicating that a number or a measurement is squared is with the superscript number two. For example, a square meter is written m², meaning that the number of meters is understood to have been multiplied by itself (that is, 100 m² would mean an area that is 10 m by 10 m). (A number used this way is called an *exponent*.)

Exploration 2.2

You're planning to put new flooring in the kitchen. The open floor in the room is 6 feet by 7 feet plus a cubbyhole for the refrigerator that measures 3 feet by 4 feet. How many pieces of floor tile will you need (assuming there is no waste) to cover all the open area if each tile measures 12 × 12 inches?

Volume means the bulk of an object, the amount of three-dimensional space it occupies. The formula for measuring the volume of a rectangular space—for example, the inside of a room—is:

$$V \text{ (volume)} = L \text{ (length)} \times W \text{ (width)} \times H \text{ (height)}$$

Your answer will be a *cubical* measurement in which the length unit used will be expressed with the exponent 3: cubic centimeters (cm³), cubic feet (ft³), and so on. The three indicates that three cumulative multiplications are involved. For example, a cubic meter (m³) represents an area 1 meter wide by 1 meter long by 1 meter high.

Exploration 2.3

The bedroom of your apartment is too hot in the summer, so you're going to. buy an air conditioner at the midwinter sale. The cooling capacity of an air conditioner is based on the number of cubic feet (ft³) of space it can cool under standard conditions, so one of the first things a salesman will ask is, "How large is your room?" Your bedroom is 10 feet by 12 feet with an 8-foot ceiling. How many cubic feet will you need to cool?

Besides the cubic footage, other factors enter into calculating cooling needs. Do you know what they would be and whether they would increase or reduce the capacity you need?

The metric unit of volume equal to 1000 cm³ has a special name: It is called the **liter** (L) and is just slightly larger in volume than the U.S. liquid quart. A *deci*liter (dL) is ¹/₁₀ of a liter (from the Latin for "ten"); a *milli*liter (mL), which is ¹/₁₀₀₀ of a liter, is the same as a cubic centimeter (cc). These measurements are commonly used in medicine and pharmacology as well as other sciences. The quart is still used in the United States in selling such items as milk and gasoline, but soda pop and other liquids are now being sold by the liter.

Weight means the pull of gravity exerted on an object by the earth or some other large body such as a planet.

Measuring Weight and Mass

What we call the **weight** of an object actually represents the pull that the earth's gravity exerts on it. In other words, an object has weight only in proportion to its nearness to another very large object— the earth, the moon, the sun, and so on. The farther from earth it moves, the less an object (for example, a space satellite) weighs. **Mass,** on the other hand, is the amount of matter an object contains. An object's mass remains constant anywhere in the universe, regardless of its nearness to earth (or a star, or another planet), provided that nothing is taken away from it or added to it. For example, two bricks together would have twice the mass of a single brick. If they were transported 1600 miles above the earth's surface, their mass would still be twice that of a single brick, but they would weigh what a single brick weighs at sea level on earth.

The basic metric standard of mass is the **kilogram** (kg), defined as the mass of a cube of platinum kept at the International Bureau of Weights and Measures. The standard substance chosen to represent the mass of a kilogram was 1000 cm³ of water. Like metric measures of length, metric measures of mass differ from each other by units of 10. Common metric mass

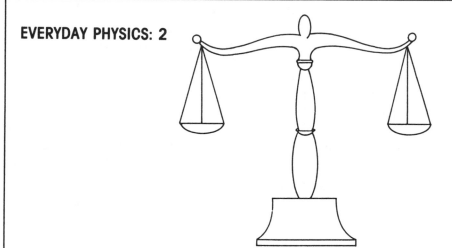

EVERYDAY PHYSICS: 2

You've probably heard the trick question, "Which weighs more, a pound of feathers or a pound of lead?" Can you see how a child could be fooled by this riddle? A feather pillow that weighs a pound looks bigger than a pound box of lead pellets, and it's common to equate size (volume) with weight (mass).

Mass refers to the quantity of matter in an object. Weight and mass are expressed by the same units.

---

units and their English equivalents are given in Table 2.2.

When we weigh an object on a balance scale, we balance it against a copy of one of the standard mass units. Basically, what we are doing is comparing the mass of the object with the mass of the standard, using the earth's gravity to do so. If we use a spring scale instead of a balance scale, comparisons are strictly valid only with other weights taken at the same height above sea level. Weighing is a convenient way of comparing masses, and that is why the same number and the same unit represent both the weight of an object and its mass.

How We Measure Time

Time is hard to define simply. It has no spatial dimensions, but you can think of **time** as a continuum along which events move from the past through the present into the future. All natural events occur in some form of time, so scientists need a standard by which to measure it. In both the English and the metric systems, the basic time unit is the **second** (sec). We measure time by the turning of the earth, and clocks are just devices for keeping step with that motion. The time of a complete turn of the earth has been divided into 24 hours. Each hour contains 60 minutes and each minute 60 seconds; there are 24 × 60 × 60 = 86,400 seconds in a day. Shorter periods than the second such as the millisecond and the nanosecond are not in general use but are used by scientists, including those in computer science.

How We Measure Density

When we say "iron is heavier than wood," we are talking about its **density:**

Table 2.2   Common Metric Units of Mass (Weight)

Metric Unit	Abbreviation	Number of Kilograms	English Equivalent
kilogram	**kg**	**primary unit**	**2.21 lb**
gram	g	0.001	0.035 oz
milligram	mg	0.000001	0.015 grain
metric ton		1000	1.10 tons
454 g			1 lb
28.4 g			1 oz

Density is the weight of a substance divided by its volume.

the weight of any object composed of a substance divided by its volume. What we really mean is that *any given volume* of iron is heavier than *the same volume* of wood. For example, a cubic foot of iron (that is, a block 1 ft × 1 ft × 1 ft) weighs 490 pounds, while a cubic foot of pine wood weighs about 30 pounds. In the English system, the density of both substances is expressed in pounds per cubic foot (lb/ft^3).

In the English system, the density of water is 62.4 lb/ft^3. In the metric system, the density of water, the standard from which the density unit is based, is 1000 grams (g) per 1000 cubic centimeters (cm^3) or, more simply, 1 gram per cubic centimeter (g/cm^3).

Table 2.3 shows the densities of some common materials in the English and metric systems. The formula for density is:

$$D \text{ (density)} = \frac{M}{V} \frac{\text{(mass)}}{\text{(divided by)}}_{\text{(volume)}}$$

By rearranging this formula, you can solve for mass or volume; in other words, if you know any two of the properties of a substance, you can find the third:

$$M \text{ (mass)} = DV \text{ (density} \times \text{volume)}$$

and

$$V \text{ (volume)} = \frac{M}{D} \frac{\text{(mass)}}{\text{(divided by)}}_{\text{(density)}}$$

Table 2.3   Densities of Several Materials

Substance	D in lb/ft^3	D in g/cm^3
aluminum	170	2.7
iron	490	7.9
lead	700	11.3
gold	1200	19.3
limestone	200	3.2
ice	57	0.92
wood, pine	30	0.5
gasoline	44	0.70
water	62.4	1.00
sea water	64	1.03
mercury	850	13.6
air*	0.08	0.0013
hydrogen*	0.0055	0.00009

*When measured at standard temperature and pressure.

For example, use Table 2.3 to find the weight (mass) of a block of ice measuring 1.5 × 1.5 × 3 feet. Note that the table gives the density of ice as 57 lb/ft³. First, find the volume ($V = LWH$):

Exploration 2.4

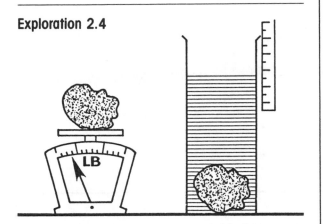

Find the density of a stone from its weight and volume. First, weigh the stone on a household scale or postal scale and record the weight in pounds. Then put some water in a straight-sided jar or glass, mark the level on the side, carefully put the stone into the water, and mark the new water level.

The volume of the stone will be the same as the volume of the displaced water, which you can visualize as a cylinder whose base is the cross-section of the jar, and whose height is the rise in water level. Measure the rise and also the inside diameter of the jar in inches.

One cubic foot = 1728 cubic inches. Your equation for finding the volume is:

$$V = \frac{\pi(\text{diameter})^2 \,(\text{height of rise})}{4 \times 1728},$$

where π (*pi*) = 3.14.

Now, use $D = \dfrac{M}{V}$ to get the density in pounds per cubic foot.

Testing Your Knowledge

2.1 Change 38.7 yards to inches.

2.2 Convert 1.34 m (meters) to inches.

2.3 Compute your height in meters.

2.4 A bolt on a French automobile has 10 threads per cm of length. How many threads per inch is this?

2.5 What is the cost of 3000 m of wire if the price is quoted as 14 cents per 100 feet?

CHAPTER 3

Force

KEY TERMS FOR THIS CHAPTER

force
force vector
resultant
equilibrium
Newton's law of gravitation

gravity
center of gravity
torque
foot-pound

Before we begin discussing force, we need to mention a point that will recur throughout this book. To make it easier to grasp physical principles, they are discussed as though they were operating in an "ideal situation," that is, as though the principle under discussion were the only one involved in a given situation. In the real world, that doesn't happen. Force, motion, and all other physical occurrences involve such variables as friction, temperature, and so on. Remember that when you study the examples in this and other chapters.

Force Vectors

A **force** can be described as the push or pull exerted on or by an object. When you use your muscles, you are exerting force. Force can be measured by weight units such as grams, pounds, or tons.

In most practical situations, we deal with not one but a number of forces acting on an object. Therefore, we need to represent each force and find the net effect of all of them. To describe a force completely, we need to specify not only its amount (say, in pounds) but also its di-

---

A force vector consists of the direction of a force plus its magnitude (amount) and can be represented on paper by a line to scale for a specified length with an arrow at the end showing direction drawn.

---

rection in space, since it makes a difference whether a force is acting to the left or the right, upward or downward. The amount (magnitude) of force plus its direction is called a **force vector.** (Any physical quantity that has both magnitude and direction is called a vector and is designated on a drawing in the same manner as force.)

To picture a force operating at a given point, physicists use a line drawn outward from that point in the appropriate direction. The length of the line is made to represent the strength of the force. For example, in Figure 3.1, line A represents a force of 5 lb that is acting toward the northeast. The drawing is made to a scale of 1 inch = 1 foot, so line A is drawn to a length of 5 quarter-inches (1¼ inches). Line B designates an eastward force of 9 lb (9 quarter-inches or 2¼ inches). The scale used is unimportant so long as it is used consistently. Any vector (for example, speed) may be drawn in the same way.

Figure 3.1

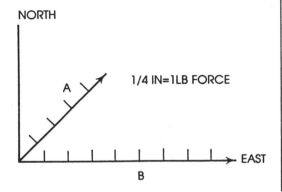

Resultant of a Set of Forces

When a number of forces act on a body, they are always equivalent to a single force of a definite magnitude and direction called their **resultant.** You can find the resultant by drawing all the forces, end to end, in any order, then drawing a line from your starting point to the end of the last force. This line will correctly represent the magnitude (amount) and direction of the resultant force.

For example, three forces act at a point: a 4-pound force acting straight down, an 11-lb force acting rightward, and a 9-lb force acting upward and leftward at a 45-degree angle. Figure 3.2a shows these forces at a scale of ⅛ inch = 1 lb of force. Now, keeping all these forces at the same length and direction, lay them end to end (Figure 3.2b). Draw a line from the starting point to the end of the last force and measure it. The line will be ¹¹/₁₆ inch long, and the resultant force, in pounds, will be 5.5 lb, or ¹¹/₁₆ multiplied by 8. Now lay the force lines down in a different order (Figure 3.2c). The resultant will be the same.

It is important to notice that the size (length) of the resultant is usually *not* equal to the sum of the magnitudes of the separate vectors. The actual resultant value will depend on their relative positions. But, if all the acting forces are in a single line (such as east-west), the magnitude of the resultant is simply the sum of all the forces acting to one side less the sum of all the forces acting toward the other (see Everyday Physics: 3).

The combined effect of all forces acting on a body can be represented by a single force of definite amount and direction called the resultant.

Figure 3.2 a,b,c

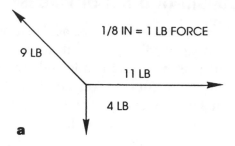

1/8 IN = 1 LB FORCE

9 LB

11 LB

4 LB

a

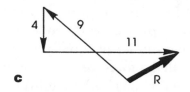

c

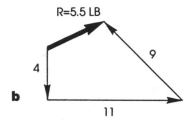

R=5.5 LB

4

9

11

b

Another situation in which the resultant can be calculated without being measured from a scale drawing occurs when two forces are at right angles (90 degrees) to each other. The resultant forms the hypotenuse (longest side) of a right triangle, and you can compute it by the

EVERYDAY PHYSICS: 3

Suppose a boy and his dad are playing tug of war. Dad can pull with a force of 100 pounds, while his son can pull only 70 pounds. If they both pull in the same direction, their combined effort will exert 170 pounds of force in that direction. If Dad pulls one way (say, west) and the boy in the opposite direction (east), the resultant will be a 30-pound force toward the west—the direction of the larger force.

rule that the hypotenuse of a right triangle equals the square root of the sum of the squares of the two adjacent sides (Figure 3.3):

R (the hypotenuse, or resultant) =
$$\sqrt{a^2 + b^2}$$

Figure 3.3

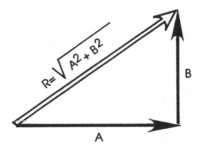

Equilibrium of Forces

Engineers and designers encounter many situations in which they must be sure that all the forces acting on a body merely hold it at rest; in other words, the resultant of forces must be zero. When this happens, the object of the forces is said to be in **equilibrium,** or balance. Looking at this from a different perspective, if an object remains at rest, we know that the resultant of all the forces acting on it must be zero. This fact can be used to find the values of some forces (see Exploration 3.1).

Exploration 3.1

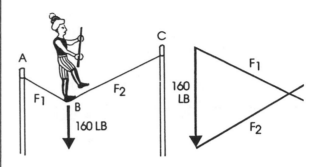

The tightrope walker is trying a walk between the New York Trade Center towers. He weighs 160 pounds. When he is at position B as shown on the drawing, what is the stretching force in each part of the cable he is walking on?

First, note that point B is the place where the forces in question meet. One of them is the man's weight. We sketch it in the downward direction from B as shown and label it "160 lb." Acting from B along the left-hand portion of the wire is some force (call it F_1) whose value is still unknown. As yet, we can only sketch it in, but we do not know how long to make it. Likewise, F_2 is the force in the other part of the wire. In most situations, F_1 and F_2 will be different.

Since the three forces hold the point B in equilibrium, they must form a *closed triangle* by themselves (zero resultant). Off to one side, draw the weight force to scale. From the tip of

---

When the resultant of forces acting on it is zero, an object is said to be in equilibrium; conversely, if an object remains at rest, we know that the resultant of forces is zero.

this force, draw a line in parallel to BC. We do not know how long to make this force; however, if we did, we would then proceed to draw the third force from its end, heading parallel to the wire AB. We would land at the starting point of the weight force. It is clear what we now have to do: Simply begin at this point and draw a line back in the proper direction until it crosses the line of F_2. This crossing point fixes the lengths (or amounts) of the two forces. Now you can measure the force lines, using the same scale you used in drawing the picture of the tightrope walker. Measuring the lines will give you the magnitudes of F_1 and F_2. Here they are about 165 lb for F_1 and 135 lb for F_2. Try a construction like this yourself, using a weight and direction of your own choosing.

Gravity

One of the greatest scientific achievements of all time was Isaac Newton's discovery of gravitation around the middle of the seventeenth century. Earlier, the astronomer Johannes Kepler had found certain regularities about the movement of planets around the sun. Newton, trying to account for these regular patterns, decided that planets must move that way because they were pulled by some force exerted by the sun. He concluded that this force, which he called gravitation, exists not only between the sun and the planets but between *any* two objects in the universe. Newton worked out the factors on which the amount of force depends and stated them as the **law of gravitation.**

Newton's law of gravitation states that any two bodies in the universe attract each other with a force that is directly proportional to their masses and inversely proportional to the square of their distance apart. That is, the greater the mass, the greater the force of attraction and the greater the distance, the weaker the force of attraction. Stated as a formula, Newton's law of gravitation reads:

$$F = \frac{Gm_1m_2}{d^2}$$

In this equation, F is the force of attraction, m_1 and m_2 are the two masses, and d is their distance apart. G is what is called a constant, and its value depends on the units chosen for F, m, and d. If F and m are measured in pounds and d in feet, the value of G is 0.000 000 000 033 $\frac{\text{ft}^2}{\text{lb}}$ (the constants are available on charts). You can see that G is very small, so the

---

Gravity increases with mass but decreases with distance; the further apart two objects are, the less the force of attraction between them. That is why astronauts experience weightlessness once they are a certain distance from the earth.

The point where an object will balance without tending to rotate in any direction is its center of gravity.

---

attraction between ordinary objects is very weak. If the objects (bodies) in question are very massive, however, as stars and planets are, the force exerted may be extremely great. For example, the attractive force between the moon and the earth amounts to about 15 million trillion pounds. The attraction of the moon for the waters of the ocean accounts for the rise and fall of the oceans known as the tides.

The gravitational force of the earth for the objects on it—what we call **gravity**—is responsible for their weight. Notice that Newton's law allows us to calculate the force of gravity but does not tell us what gravitation is or why it exists. These are questions of philosophy, not physics.

Center of Gravity

In everyday experience, the forces acting on an object are applied at several different places, not all at one point. For example, the earth's gravity pulls downward on every particle of a material object with a force equal to the weight of that particle. However, we can replace all those separate forces by a single force equal to the entire weight of the object and considered to be acting at a given place called the **center of gravity.**

For an object composed of a single material and having a simple shape—an iron cannonball, a brick, a straight rod, for example—the center of gravity and the center of the object will be the same. For an irregular object, the center of gravity may

be found by trying to balance the object. The point at which the object will balance without tending to rotate in any direction is its center of gravity. If a body is supported at any point other than its center of gravity, it will try to move until its center of gravity is as low as possible. This explains why you can't balance a pencil on its point.

Exploration 3.2

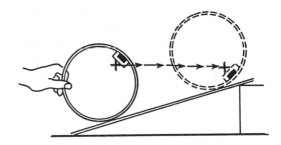

Have you ever seen anything rolling *up* a slope? You can make this happen. First, clean out an empty soup can or an oatmeal box and fully remove both the top and the bottom. Now you have a cylinder. Tape a quarter or some other small weight inside it. Prop up one end of a breadboard or some other flat board. Place your cylinder at the low end of the board with the quarter in the two-o'clock position (see sketch). When you release the cylinder, it will roll *up* the hill. Why? Because its center of gravity is very near the position of the quarter. The center of gravity will go *down*, causing the cylinder to roll up the hill.

A unit called a foot-pound is commonly used to express torque.

Torque

Unless all the forces applied to an object are acting at a single point, it will tend to rotate. The ability of a force to produce rotation is called its turning effect, or **torque.** The torque of any force is equal to its amount multiplied by its distance from the pivot point to the line of the force. To understand the idea of a pivot point, picture a revolving door. To turn the door most effectively, you will push it at its edge (the farthest distance from its hinge, or pivot point) rather than near the hinge (Figure 3.4). The distance from your hand to the pivot point is called the torque arm.

Figure 3.4

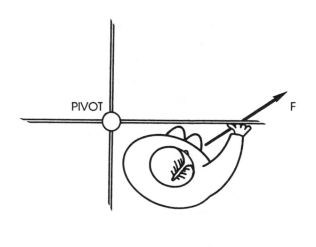

PIVOT   F

To find the torque, you use the formula $T = Fh$, in which T is the torque, F is the force, and h is the torque arm stated in feet. You do not need an actual pivot or axle in order to compute torque; you can designate any point as the prospective point of turning. When you find the torque, it will be expressed in a unit called a **foot-pound.**

In order for a body not to rotate, the net torque must be zero. That is, the sum of all torques that tend to turn the body clockwise must be equal to the sum of all those that tend to turn it counterclockwise.

Now, consider an example of the kind of force computation engineers commonly encounter: figuring out the downward force exerted by a truck on a bridge (Exploration 3.3).

Exploration 3.3

A 5-ton truck stands 30 feet from one pier of a uniform bridge 100 feet long weighing 20 tons. Find the downward force on each pier.

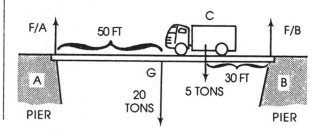

For an object not to rotate, net torque must be zero; that is, the sum of all clockwise forces must equal all counterclockwise forces.

First, you need to note all the forces acting *on* the bridge:

A 5-ton downward force at *C*.

A 20-ton downward force at *G*, the center of gravity of the bridge structure.

At the piers, upward forces F_A and F_B whose values are to be found.

Take torques around *A*. The two weight forces tend to turn the bridge clockwise about *A*, and their torques amount to 20 × 50 + 5 × 70, or 1350 foot-tons. The only counterclockwise torque is that of F_B, amounting to 100 F_B. Notice that F_A does not contribute any torque, since it has no torque arm around *A*.

Setting the torques in the two directions equal, 100 F_B = 1350 and F_B = 13.5 tons force.

You could now repeat the process, taking torques around, say, the point *B*; but there is a simpler way to find the remaining force F_A. You know that for the bridge to be stable, that is, at equilibrium, the resultant of all the acting forces must be zero. Simply because all the forces in this problem are either upward or downward, we have F_A + 13.5 = 20 + 5, so that F_A = 11.5 tons. So we see that by using the two equilibrium conditions that state (1) the resultant of all the forces must be zero and (2) the torques around any point must balance, we can work out any equilibrium problem.

Testing Your Knowledge

3.1 A force acting at a point may be completely described by stating
    a. its amount in pounds.
    b. all the other forces that act.
    c. its amount and direction.
    d. its direction in space.

3.2 The resultant of a number of forces acting at a point is (choose one)
    a. the single force that produces the same effect.
    b. not fixed in direction, but has a definite magnitude.
    c. dependent on the order in which the forces are taken.
    d. zero under all circumstances.

3.3 A body is said to be in equilibrium if all the forces acting on it
    a. have the same direction.
    b. are of equal magnitude.
    c. have zero resultant.
    d. are arranged in opposite pairs.

3.4 A block of wood resting on a table is pulled by two cords attached to it. One exerts a force of 100 g, the other a force of 50 g. In order to get the

greatest net force on the body, the two cords should be pulled
a. in opposite directions.
b. at right angles to each other.
c. in nearly opposite directions.
d. in the same direction.

3.5 A wire is pulled tight at a right angle around the corner post of a fence, there being a force of 50 lb in each part of the wire. The resultant pull on the post (make a drawing to scale) will be
a. less than 50 lb.
b. directed halfway between the two wires.
c. greater than 100 lb.
d. exactly 100 lb.

3.6 In terms of the idea of center of gravity, explain why a ship is less stable when empty than when loaded.

3.7 Two men carry a 150-lb load hung from a lightweight pole resting on their shoulders. If the load is attached at a point 4 ft from one man and 5 ft from the other, how much of the weight does each carry?

Hint: Follow the method of Exploration 3.3.

3.8 A uniform brick whose dimensions are $8\frac{1}{2} \times 4 \times 2$ inches weighs 5.4 lb. It rests with its largest face on the floor. How big an upward force, applied at the center of the smallest face, will just lift one end off the floor?

3.9 If the earth were three times as far from the sun as it is now, how would the gravitational attraction compare with its present value?

3.10 Compute the force of attraction between two 15,000-ton ships whose centers of gravity are effectively 150 ft apart.

CHAPTER 4

Motion

KEY TERMS FOR THIS CHAPTER

speed
velocity
acceleration
acceleration due to gravity
law of inertia (Newton's first law)
centripetal force

centrifugal force
law of constant acceleration (Newton's second law)
law of conservation of momentum (Newton's third law)

In the world around us, everything moves—even objects at rest upon the ground. Objects at rest are really moving with the rotation of the earth, which is also moving on its path around the sun, and so on. In other words, rest and motion are relative terms. In this chapter we discuss how to measure the motion of objects and how the forces acting on objects determine the way they move.

Speed and Velocity

When we move or watch something moving, we are interested in two things

about its motion: its rate and its direction. **Speed** is the rate at which something moves; it is measured by the distance covered divided by the elapsed time. In symbols:

$$v\,(\text{speed}) = \frac{d\,(\text{distance})}{t\,(\text{time})}\,(\text{divided by})$$

Even if the rate of speed is not constant over the entire distance, this formula is useful in giving the average speed for the trip. For instance, if you drive to a city 150 miles away in a total time of 3 hours, your average speed will be 150 miles divided by 3 hours, or 50 miles per hour.

Table 4.1   Conversion Factors for Common Speed Units

	mi/hr	ft/sec	cm/sec	knots
mi/hr	—	1.47	44.7	0.868
ft/sec	0.682	—	30.5	0.592
cm/sec	0.0224	0.0328	—	0.0194
knots	1.15	1.69	51.5	—

Speed is a derived unit, and any distance and any time unit can be used to determine and express it. Common speed units include miles per hour (mph), feet per second (ft/sec), centimeters per second (cm/sec), and knots per hour. (A *knot* is a nautical mile equal to 1.15 land miles; the speed of ships, and sometimes airplanes, is measured in knots.) Table 4.1 gives factors for converting some common speed units into others. To change a unit listed at the side to one listed at the top, multiply by the factor in the appropriate cell (square).

Velocity is speed plus direction. Like force (see Chapter 3), speed is a vector and can be represented graphically by a segmented, arrowed line. An object can have several velocities at the same time. For instance, Figure 4.1 is a schematic top view representing a ball being rolled across a moving railroad car. The ball shares a common velocity with the train and everything in it, plus the crosswise velocity at which it is rolled. The doubled arrow in Figure 4.1 indicates the resultant velocity of the ball; since the two vectors form a right triangle, this velocity can be computed by the formula

$$v = \sqrt{a^2 + b^2}$$

Figure 4.1

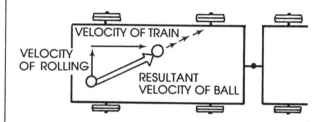

Acceleration

Most of the motion we observe does not occur at a constant speed, whether it is the flight of a bird or the falling of a stone. Any motion in which speed or direction varies is called accelerated motion. **Acceleration** is defined as the rate of change of the velocity; that is, the change in velocity divided by the time it takes for the change to occur.

Acceleration is the rate of change in velocity, expressed by the distance unit squared—for example, ft/sec² or mph².

To find out how far a constantly accelerating object can travel in a given time, multiply its average speed by the time traveled.

For instance, suppose a car going 25 ft/sec picks up speed until, 5 seconds later, it is going 60 ft/sec. Its rate of pickup will be 60 − 25, or 35 ft/sec in 5 seconds. Dividing, this amounts to 7 feet per second per second (7 ft/sec/sec), meaning that in each second it was accelerating, the car increased its speed at the average rate of 7 ft/sec. Notice that the time factor occurs twice as a factor in the derived unit, meaning that it can be written as ft/sec².

If the object is slowing down instead of speeding up, that is, if it is decelerating, the rate is shown with a minus sign in front of it.

Constant acceleration is easily described and computed. This kind of acceleration (or deceleration) may occur for a limited time when an object is gathering speed or is stopping. For example, suppose a car traveling at 30 ft/sec is brought to rest by its brakes at the uniform rate of 5 ft/sec². How long must the brakes be applied? By saying that the braking acceleration (this is acceleration of the braking action but deceleration of the car) is − 5 ft/sec², we are saying that the car will lose speed at the rate of 5 ft/sec each second. Therefore, at 30 ft/sec divided by 5, the car will have lost all its initial speed at 6 seconds.

How far will a constantly accelerating object move in a given time? Remember that the speed of motion is changing all the while. Here you can make use of the average speed. Since the speed is changing at a uniform rate, the average speed will be the speed at the beginning plus the speed at the end, divided by two (see Exploration 4.1).

Exploration 4.1

Suppose a car going 26 ft/sec begins to accelerate at the rate of 2 ft/sec². How fast will it be going after 8 sec, and how far will it go in this time? In 8 sec, the total gain in speed will be 8 × 2 = 16 ft/sec, so the final speed will be 26 + 16, or 42 ft/sec. To find the distance traveled, we note that the speed at the beginning of the acceleration period was 26 and at the end was 42 ft/sec, so that the average speed over this interval is ½ (26 + 42) = 34 ft/sec. Going, in effect, 34 ft/sec for 8 sec, the car would cover a distance of 34 × 8, or 272 ft.

Now, what about objects that are falling? At one time, it was thought that heavier objects fall faster than lighter ones. However, the experiments conducted by the great Italian scientist Galileo in the sixteenth century showed that if one ignores the effect of the surrounding air, falling bodies have constant acceleration. This is called **acceleration due to gravity** and is denoted by the symbol G. The value of G changes slightly from place to place on earth and especially with height, but the standard value of G is close to 32 ft/sec or 980 cm/sec.

If you know the value of G, you can calculate the motion of any falling object (see

Exploration 4.2). Your results will be quite accurate for compact solid objects falling for short distances. For bodies falling great distances through the atmosphere, G cannot be so readily calculated.

You can use the standard value of G to calculate not only the speed of a falling object but the distance it fell. For example, you see a flowerpot falling from the ledge of an apartment balcony. It hits the ground 7 seconds later. How high is the ledge, and how fast was the flowerpot falling? At a constant acceleration, in 7 seconds the flowerpot, picking up speed as it falls, would be traveling at 7 (sec) × 32 (ft/sec²) = 224 ft/sec when it hit the ground. Its *average* speed would be 0 ft/sec (speed at the beginning) plus 224 ft/sec (speed at end) divided by 2, or 112 ft/sec. Going at this speed for 7 seconds, the flowerpot would have covered a distance of 112 × 7 = 784 feet, which is the height of the balcony.

A *projectile* such as a thrown shotput or a bullet is really a falling body. If shot upward at an angle (Figure 4.2), it immediately begins to fall short of the direction of fire. It continues to fall while moving forward, and so it follows a curved path. In the case of bullets, which travel at high speed, the path may be somewhat altered by air resistance.

Figure 4.2

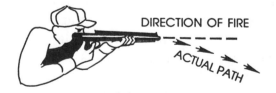

DIRECTION OF FIRE

ACTUAL PATH

Exploration 4.2

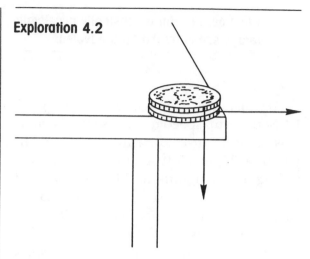

Place two coins at the very edge of a table, one on top of the other. A sharp blow with a knife blade held flat against the table will send the lower coin off like a projectile, while the upper one will fall almost straight down. In spite of this difference in path, you will hear both strike the floor at the same time, since both really *fall* the same distance.

Inertia

We have now described certain types of motion and calculated speeds and distances. But what causes objects to start moving? Once they are in motion, what keeps them moving? These questions occurred to Isaac Newton also, and the laws he formulated to answer them remain valid in everyday physics centuries after he did so. These laws, called Newton's laws of motion, include the law of inertia (Newton's first law of motion), the law of constant acceleration (Newton's second law of motion), and the law of momentum (Newton's third law of motion).

The **law of inertia** describes a fundamental property of matter: Every body (object) remains in a state of rest or of uniform motion in a straight line unless acted upon by outside forces. This law states that motion is as natural a condition as rest. Just as an object at rest is in equilibrium, so is an object moving in a straight line at a constant speed. For example, a car that is going along a level road at constant speed is balanced by the supporting forces of the pavement, and the forward pull of the engine counterbalances the retarding forces of friction and air resistance. The resultant force is zero; thus the car is in equilibrium.

If the car comes to a curve, the pavement must furnish, through friction with the tires, an additional force to swerve the car from its natural straight path so that it can round the curve. If the road is slippery, so that friction is lacking, the car will continue straight ahead, tending to skid off the road.

Figure 4.3

The force required to hold a moving object in a circular path is called **centripetal** (toward the center) **force.** Properly engineered curves in a road or the curves on a bicycle racetrack are banked, or raised at the outer edge to furnish such a force. A satellite following an orbit around a planet or a planet going around the sun is held in orbit by the centripetal force furnished by gravitational attraction.

Conversely, **centrifugal** (away from the center) **force** causes objects to fly off on a tangent. For example, mud flying from a spinning car wheel moves in a straight line away from the center of rotation. A washing machine on "Spin" uses centrifugal force to whirl the clothes against the sides of the basket. A device called a centrifuge is used by chemists and biologists to separate suspended solid matter from liquid. The difference in the amount of centrifugal force exerted on the solid material and the less dense liquid causes the solids to collect at the outer rim.

Constant Acceleration

Newton's first law, the law of inertia, tells us what happens to objects in motion and at rest when the resultant force is zero; that is, when there is no resultant force. In most real-life situations, outside forces do act, and Newton's second law helps predict what will happen when they do. Newton's **law of constant acceleration** states that a body acted upon by a constant force will move with constant acceleration in the direction of the force; the amount of acceleration will be directly proportional to the acting force and inversely proportional to the mass of the body.

The law of constant acceleration (Newton's second law) tells us that a body acted upon by a constant force will move with constant acceleration; acceleration increases as force increases and decreases as mass increases.

Put into a form that allows computations, Newton's second law can be stated as:

$$\frac{F \text{ (force)}}{W \text{ (weight)}} \text{ (divided by)} = \frac{a \text{ (acceleration)}}{g \text{ (gravity)}} \text{ (divided by)}$$

This law can also be expressed as $F = ma$.

Remember what happens when an object falls, that is, accelerates by gravity. Here, the force is *any* applied force that is equal to the weight of the object, and the acceleration is that of gravity. Any weight unit can be used for F and W, and any acceleration unit (such as ft/sec²) can be used for a and g. For example, a car weighing 3200 lb accelerates at the rate of 5 ft/sec². Ignoring friction, what is the effective forward force exerted by the engine? W equals 3200 (the weight of the car); a equals 5 (the acceleration of the car) and g equals 32 (the acceleration of gravity). Substituting numbers:

$$3200 \times {}^5/_{32} = 500 \text{ lb force}$$

This law explains why pilots and astronauts experience what you've probably heard of as G forces. The weight of a person at rest is the force exerted by and acceleration of gravity, 1 G, or 32 ft/sec². If someone is being accelerated at a rate greater than 1 G, he will feel heavier. For example, if he is accelerated at 64 ft/sec², he will feel as though his weight had doubled. An easy way to say this is that he is feeling a 2 G ($2 \times 32 = 64$) force. If accelerated at 96 ft/sec², he will feel a 3 G force, and so on.

Momentum

Newton's third law deals with a situation you may have heard as "for each action, there is an equal and opposite reaction." It means that it is not possible to exert a force on an object without exerting a force in the opposite direction on some other object (body) or objects. If you jump from a rowboat to a pier, the force of your jumping pushes the boat backward. A gun "kicks" when the bullet goes forward. A ship's propeller can drive it forward only because it continually throws water backward.

Newton defined the *momentum*, or propulsive force of a body, as mass multiplied by linear velocity:

$$M \text{ (momentum)} = m \text{ (mass) times } v \text{ (velocity)}$$
$$\text{or}$$
$$M = mv$$

Momentum, or the propulsive force of an object, equals the object's mass multiplied by its velocity ($M = mv$).

M is a derived quantity, and any appropriate units may be used for *m* and *v*. Newton's third law, often called the **law of conservation of momentum,** states that, when any object (body) is given a certain momentum in a given direction, some other body or bodies will receive an equal momentum in the opposite direction.

Exploration 4.3

Suppose that a gun has a mass of 2500 grams and the bullets each have a mass of 100 grams. If a bullet leaves the gun with a speed of 800 meters/sec, with what speed will the gun start back? The momentum of the bullet will be 100 × 800 gm m/sec (gram meters per second). Calling the recoil speed of the gun *V*, its momentum just after firing will be 2500*V*. Setting the two momenta equal, 2500*V* = 100 × 800, so that *V* = 32 m/sec. *V* comes out in m/sec because the speed of the bullet was given in these units.

If the gun and bullet were subject to no other forces after firing, the two would go in opposite directions, each continuing to move with its own constant speed forever (first law). This would nearly be the case, for example, if the gun were fired far out in space where friction and gravitational forces are negligible. If the gun were fixed in the ground rather than free to recoil, the reaction would be transmitted to the whole earth instead of to the gun alone. Because of the earth's enormous mass, its resulting motion would be far too small to be detectable.

Exploration 4.4

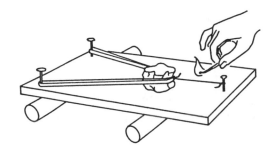

You can demonstrate the reaction principle by making a rubber-band slingshot on a board resting on rollers. Tie the band back with a string and place a fairly massive stone in firing position. Release the stretched band by burning the thread and observe the recoil of the board as the stone goes forward.

Newton's laws apply to rotation as well as to forward motion. An object (body) that is set spinning has a tendency to keep spinning. The heavy flywheel of an engine illustrates a use of this principle to smooth out the separate power thrusts created by the explosion of fuel and the resultant movement of the pistons.

A rotating wheel tends to keep its axis in a constant direction in space, as illustrated by a spinning top. The top remains vertical so long as it spins fast enough. This is the principle of the gyroscope, a rapidly rotating wheel mounted in a pivoted frame so that the axis may hold its direction despite any motion of the mounting. This ability to keep its direction makes the gyroscope useful in aircraft instruments such as the turn indicator, artificial horizon, gyrocompass, and automatic pilot.

Inertial guidance systems used in missiles usually contain three gyroscopes to provide fixed references of direction on three axes.

EVERYDAY PHYSICS: 4

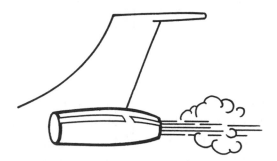

A jet engine or a rocket gets its propelling force from the reaction of the gases discharged toward the rear at high speed. The mass of gas shot out each second is not very large, but its great velocity makes the momentum (the product mv) very great. The plane or rocket gets a momentum in the forward direction equal to the rearward momentum of the discharged gases.

Testing Your Knowledge

4.1 A delivery truck covers 2 miles of its route at a speed of 24 mi/hr, makes a stop for 15 min, and then takes 6 min to go the remaining $1\frac{1}{2}$ mi. What is the average speed for the whole trip?

4.2 If the acceleration due to gravity on the moon is $\frac{1}{6}$ the value on earth, how fast will a freely falling stone be moving on the moon 2 sec after it is dropped?

4.3 Compare the distance a stone falls in the first second after being dropped with the distance it falls during the second second.

4.4 If the stone in the preceding example were *thrown* downward instead of being dropped from rest, how would this affect its average speed for the whole trip? How would this affect the time required to fall the whole distance?

4.5 Considering Figure 4.2, must one aim high or low in order to hit a distant target? Explain.

4.6 Collisions usually involve very sudden changes in speed or direction of motion on the part of the bodies involved. How does this explain the destructiveness of an automobile collision at high speed?

4.7 Explain how it is possible to exert a blow of several hundred pounds of force by means of a hammer weighing only a few pounds.

4.8 If the gun and bullet get equal amounts of momentum, why is it less dangerous to take the "kick" of the gun than to be hit by the bullet?

4.9 In walking, we push back on the ground; the reaction is the ground pushing forward on us. Why is it difficult to walk on ice?

4.10 Why is it harder to stop a ferry boat than a canoe going at the same speed?

CHAPTER 5

Work and Energy

KEY TERMS FOR THIS CHAPTER

mechanics

work

energy

mechanical energy

potential energy

gravitational potential energy

kinetic energy

pendulum

power

horsepower

watt

machine

What Is Work?

What does "work" mean to you? Probably it means your job. Or you may think of it in terms of effort. If you come in from the garden after a battle with a stubborn boulder, you wipe the sweat from your forehead and head for a cold drink as a reward for all that work. However, if you didn't manage to move the boulder, you expended energy but did no work in the sense in which physicists use the word.

In **mechanics,** the branch of physics that analyzes the action of forces on matter, **work** occurs only when a force succeeds in moving the body (object) it acts on (see Exploration 5.1). The amount of work done is equal to the amount of force multiplied by the distance the object moves in the direction the force is acting in. In other words,

W (work) = f (force) times d (distance)
or $W = fd$

Exploration 5.1

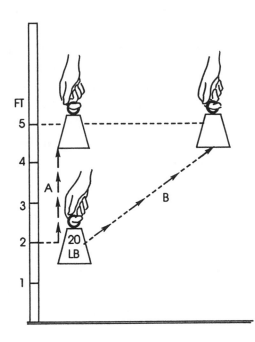

Suppose you are holding a 20-lb dumbbell 2 feet from the floor. How much work are you doing? Then, how much work are you doing if you lift it straight up from that point to a height of 5 feet from the floor? How much additional work would you be doing if you lifted the dumbbell in zigzag fashion?

First, simply holding the dumbbell accomplishes *no* work, because no movement is involved. Second, lifting this 20-lb weight slowly and steadily to a height of 5 feet requires an upward

force of 20 lb, moving through a distance of 3 feet in the direction of the force. The work done is 20 × 3 = 60 ft-lb.

Now, remember that the only thing that matters in calculating work is the *distance moved in the direction of the force.* When you move the weight in zigzag fashion, you are moving it sidewise as well as upward; in terms of work, however, only the vertical movement counts.

Work is a derived unit; it may be expressed in any force unit times any distance unit such as foot-pounds (ft-lb), kilogram meters (kgm), and so on. A metric work unit called the *erg* expresses work in centimeters per gram per second (cmg/sec²). It is used in computing very small forces. A *joule*, which is more commonly used, equals 10 million ergs and is almost equal to 0.75 ft-lb. It is named after James Prescott Joule (1818–89), an English physicist.

Energy

Physics does not deal only with matter; it also deals with such things as electricity, light, sound, and heat. These are not kinds of matter; they do not take up space nor have weight. They are forms of **energy,** and their activity produces changes in matter. For example, heat can change a liquid to a gas. Light can form an image

---

In physics, the term work does not apply to the mere expenditure of effort; work occurs only when a force succeeds in moving the object it acts on.

In classical physics, energy is not considered material; instead it produces changes in matter. Light, heat, and electricity are forms of energy.

on a photographic film by changing the state of the film's silver coating.

Electrical energy can turn a motor, plate a spoon with silver, transfer your ideas to a magnetized disk, or send your voice over thousands of miles of space. Chemical energy heats your home directly (through the burning of coal, wood, oil, or gas) or indirectly (by electricity generated by burning a fuel). The potentially destructive power of atomic energy is widely known through pictures of Hiroshima and Nagasaki after they were bombed in World War II.

Probably the most familiar energy effects are those that make bodies move to change their motion. **Mechanical energy** has been called the "go" of things. The mechanical energy of any object (body) is measured by the amount of work it can do. Total mechanical energy generally consists of several different kinds of mechanical energy.

Potential energy, or energy of position (abbreviated PE), represents the amount of work an object can do by returning to its original position. For example, opening a jack-in-the-box releases the potential energy of the coiled spring inside. When you wind a watch, you do work in coiling its spring tighter; this work is stored in the spring and slowly released to run the watch mechanism. A stick of dynamite has chemical potential energy; so does a lump of coal or the battery you use in your cassette player.

A raised weight possesses potential energy in a form called **gravitational potential energy** (GPE). When it is released to return to its former level, it can do work such as raising another weight or stretching a spring.

Have you seen a pile driver at work? The weight is raised and then dropped. As it falls, it acquires speed, and by the time it hits the piling, it has enough energy to drive the piling in. How did it acquire that energy? The potential energy stored in it as it was raised was converted to energy in the form of motion. The energy of a moving body is called **kinetic energy** (KE).

In the course of the pile driver's operation, several different kinds of energy are involved. At the point of release, the falling weight had only potential energy. Just before hitting the piling, it had only kinetic energy. It acquired this kinetic energy expending its potential energy.

Ignoring friction, air resistance, and other factors that would be present in a real situation, we can say that the total mechanical energy, consisting of PE + KE, remains constant. As KE is lost, PE is gained. The principle that the total mechanical energy of any system remains constant is called the principle of *conservation of mechanical energy.*

Potential energy (PE) is energy of position; this energy is released when an object returns to its original position. Kinetic energy (KE) is energy of motion.

Exploration 5.2

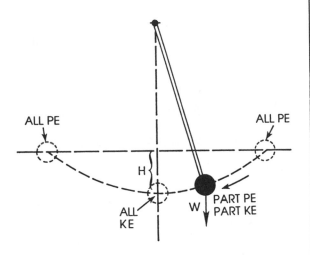

A **pendulum** is another example of conversion between gravitational potential energy and kinetic energy. When it is first pulled up, before being allowed to swing on its own, work is done against gravity because the bob is effectively raised to a distance *h*. At the position *A* it possesses an amount *Wh* of potential energy. If released, the pendulum swings down. As it moves toward the bottom of its arc, its gravitational potential energy becomes kinetic energy. As it reaches the bottom point, all its energy is kinetic. Then, if we ignore air resistance, the bob climbs up to the original level, its energy changing from kinetic to potential. However, in real life, air resistance and friction do exist, and some energy would be lost with each swing. The pendulum cannot swing perpetually without more work being done.

The kinetic energy of a moving body can be calculated multiplying its weight by the square of its velocity and dividing by 64 (2 G). If the weight is in pounds and velocity is in feet per second, KE will be expressed in foot-pounds:

$$KE\ (\text{ft-lb}) = \frac{m\ (lb)v^2(\text{ft}^2/\text{sec}^2)}{2 \times 32\ (\text{ft}/\text{sec}^2)} = \frac{mv^2}{64}$$

If mass is in grams (gm) and velocity is in centimeters per second (cm/sec), then KE will be given in gm-cm by:

$$KE\ (\text{gm-}\frac{\text{cm}^2}{\text{sec}^2}) = \frac{m\ (gm)\ v^2\ (\text{cm}^2/\text{sec}^2)}{2} = \frac{mv^2}{2}$$

For example, how much kinetic energy is possessed by a 3000-lb car moving at 60 mph? Keeping in mind that 60 mph equals 88 ft/sec, *KE* = 3000 × (88)²/64 = 363,000 ft-lb.

Or, suppose a 2-lb bucket of paint drops from a 9-ft ladder. Ignoring air resistance, what is its KE and speed just before it strikes the ground? To raise the paint bucket to the top of the ladder required work in the amount of *W* = 2 × 9 = 18 ft-lb, and this is the amount of its original GPE. By the time the bucket reaches the ground, all the PE has become KE, so the magnitude of KE is also 18 ft-lb. To find the speed, using 2 lb for *m:*

$$2v^2/64 = 18$$
$$v^2 = 576$$
$$v = 24\ \text{ft/sec}$$

In the absence of other forces that dissipate energy, such as friction or air resistance, the total mechanical energy of a system remains constant; this is called the principle of conservation of mechanical energy.

Power

In many practical applications, we want to know how long it takes to do a certain piece of work. In winding a watch, storing elastic energy in the spring may take only about 10 seconds, while the same amount of energy may be released over perhaps 30 hours as the watch runs down. The chemical energy stored in a tree by sunlight over 50 years may be released in 50 minutes when the wood is burned.

The rate of doing work is called **power**, which is work divided by time, or:

$$P\,(\text{power}) = \frac{W}{t}\begin{array}{l}(\text{work})\\(\text{divided by})\\(\text{time})\end{array}$$

One well-known power unit is **horsepower** (hp). James Watt, who made the steam engine practical, measured the rate at which a horse could work and found it to be about 550 ft-lb/sec. For example, a horse should be able to raise a 275-lb weight at the rate of 2 ft/sec. This standard unit of power can be written in symbols as:

$$P\,(\text{hp}) = \frac{W\,(\text{ft-lb})}{550 \times t\,(\text{sec})}$$

One of the metric units of power has a special name; it was named for the same James Watt who devised the unit horsepower. A **watt** is a working rate of 1 joule per second; a kilowatt (kw), which is used especially by electrical engineers, is equal to 1000 watts. One hp is equal to about 0.75 kw.

Suppose a 3000-lb car moving at 60 mph attained its full speed in 15 sec from a

EVERYDAY PHYSICS: 5

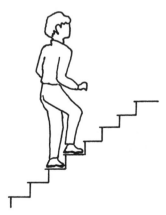

A horse can presumably go on expending energy at about 1 hp for extended periods of time. A human being can exert surprisingly large amounts of power but only for short intervals. Determine your power output in running up a flight of stairs. You will have to know your weight, the height of the stairs (vertical distance) and the time. If possible, use a stop-watch for timing yourself.

standing start. Knowing from an earlier example that its KE is 363,000 ft-lb, what power has the engine exerted?

$$P(\text{hp}) = \frac{363,000}{550 \times 15} = 44 \text{ hp}$$

Machines

You may think of a machine as something large and noisy, with an engine and a lot of moving parts—a bulldozer, for example. In physics, however, a **machine** is any device by which energy can be transferred from one place to another or one form to another. Have you used a stick to pry up a seashell buried in the sand? The stick was a machine.

When a machine is used, some outside agency—a motor, a battery, your muscles—does work on the machine. The machine then delivers work to something on which it acts. The principle of conservation of mechanical energy dictates how these two kinds of work are related: So long as any energy stored up in the machine itself remains constant, and in the absence of any friction, the work done *by* the machine is exactly equal to the work done *on* it.

In the real world, however, friction and other energy-wasting factors such as air resistance are never absent. This means, among other things, that no machine can keep itself going forever; in the real world, one cannot build a perpetual-motion machine.

Testing Your Knowledge

5.1 When a block of wood acted upon by a force of 48 lb moves 10 ft in the direction of the force, the work done is
  a. 4.8 ft-lb.
  b. 480 ft-lb.
  c. 58 ft-lb.
  d. 0.21 ft-lb.

5.2 Of the following, all are possible units of mechanical energy except
  a. ft-sec.
  b. ergs.
  c. joules.
  d. in-lb.

5.3 Energy is
  a. work divided by time.
  b. the ability to do work.
  c. measurable in horsepower.
  d. force divided by distance.

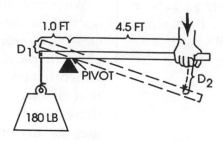

5.4 *Principle of the lever.* In the arrangement shown, how large a downward push is needed to raise the load, friction being negligible?
   Hint: Use the work principle and notice that the distances moved by the applied force and the weight force are directly proportional to the distances from the pivot.

5.5 A man rolls a 200-lb keg up a sloping board to a loading platform 4.0 ft high. The board is 10 ft long, and rolling friction can be neglected. How hard must the man push if the force he exerts is parallel to the incline at all times?
   Hint: The amount of work required will be the same as that needed to raise the keg straight up through a distance equal to the height of the platform.

5.6 When a lever (crowbar) is used under the conditions specified in problem 5.4, how does the *power* exerted by the applied force compare with the power expended on the load?

CHAPTER 6

Heat

KEY TERMS FOR THIS CHAPTER

temperature
ice point (freezing point)
steam point (boiling point)
degree
Celsius scale
Fahrenheit scale
absolute zero

kelvins (degrees Kelvin)
coefficient of linear expansion
conduction
coefficient of heat conduction
convection
radiation
greenhouse effect

Heat is necessary to life. It is also one of our most valuable tools: It cooks our food, frees metals from their ores, refines petroleum, and runs trains and automobiles. Not so long ago, heat was thought to be an invisible, weightless substance that passed from a hot to a cold object. Today, we know that heat is a form of energy: the energy created by the motion of molecules.

Temperature and How We Measure It

Temperature is the degree of hotness or coldness of an object or an environment. We perceive temperature by means of special nerve endings in our skin, but our judgment of hot and cold may be influenced by many other factors (see Exploration 6.1).

Nearly all materials expand when their temperature is raised and shrink when it is lowered.

Exploration 6.1

Place three bowls in a row—the first containing cold water, the second lukewarm water, the third hot water. Put your left hand in the cold water, your right hand in the hot. After a few seconds, remove and plunge them at once into the middle bowl. The lukewarm water in it will seem hot to your left hand and, at the same time, cold to your right, although you know it to be the same temperature all through.

Generally, when the temperature of matter is changed, other things happen to it also. Its size or its electrical, magnetic, or optical behavior may change, and any such change might be used to detect and measure its change in temperature. However, in most cases, the simplest change to use is change in size. Nearly all materials expand when their temperature is raised and shrink when it is lowered.

The common mercury thermometer makes use of the expansion and contraction of liquid mercury to measure temperature. The mercury's very slight changes in bulk are made more evident by attaching a fine tube to the bulb. To specify a scale by which changes can be measured, two fixed points are chosen: the **ice point** (also known as the freezing point) and the **steam point** (also known as the boiling point).

The position of a mercury thread on any given thermometer placed in a mixture of ice and water is always the same; this is the ice point. Again, when the instrument is held in the steam rising from boiling water it always comes to a fixed level; this is the steam point. The distance between these two points is then marked off in equal increments called **degrees.**

In the **Celsius system,** which is used in scientific work, the ice point is designated as 0 degrees (0°C), and the steam point is designated as 100 degrees (100°C). The space between is divided into 100 equal parts, each being 1 degree Celsius (1°C). The Celsius system, which is named in honor of Swedish astronomer Anders Celsius, was formerly called the centigrade system (*centi-* is Latin for "hundred").

In the **Fahrenheit scale,** which is still in use, the fixed points were originally based on the temperature of a mixture of ice and salt and the temperature of the human body. In the Fahrenheit scale, the ice point is at 32 degrees (32°F) and the steam point at 212 degrees (212°F). To convert either reading to the other scale, you use the formula:

$$(\text{degrees}) \, F = \frac{9}{5} (\text{degrees}) \, C + 32$$

except that, if the temperature on either scale is below zero, you must place a minus sign in front of its number in the equation. For example, the temperature of

---

The ice point, or freezing point, and the steam point, or boiling point, are the reference points used in constructing a scale for measuring temperature.

Kelvins are measurements of temperature based on Celsius degrees starting from absolute zero, the point at which molecular motion ceases; the ice point is 273 and the steam point is 373.

---

solid carbon dioxide ("dry ice") is about −80°C. To find its temperature in degrees Fahrenheit:

$$°F = \frac{9}{5}(-80) + 32$$
$$= -144 + 32 = -112°F$$

Today, many thermometers give readings in both scales.

Another scale of temperature used in theoretical science dealing with molecules has been called the *Kelvin* or *absolute scale.* This scale uses the Celsius degree but with a starting point of **absolute zero,** the point at which gases cease to exert pressure (see Expansion of Gases below).

Absolute zero was formerly considered to be the lowest temperature possible in the universe. (In principle, there is no upper limit to temperature.) In this scale, there are no negative temperatures (below zero). The ice point is 273 degrees and the steam point is 373 degrees. The increments in this system are now simply called **kelvins;** a reading in kelvins can be converted to degrees Celsius simply by adding 273. For example, a summer day's temperature of 27°C would be 300 kelvins (27 + 273).

Expansion and Contraction of Solids and Liquids

Warming a substance means giving more energy to its molecules; cooling it takes energy away. Most substances expand when they are warmed, but they do so at different rates. For example, an iron bar 1 foot long at ice point increases in length about 1/70 inch when heated to steam point. A brass bar expands about 1.5 times as much, while a glass rod expands only half as much. It has been found that solid objects composed of different substances increase in length by a certain fraction for each rise in temperature; this fraction, which can be converted to either degrees Fahrenheit or degrees Celsius, is called the **coefficient of linear expansion.**

For example, the coefficient of expansion of iron or steel is 0.000011. Compare this with the coefficient for the alloy invar (an alloy is a mixture of metals), which is 0.0000009. When designing machines and structures, engineers must carefully consider the comparative coefficients of expansion of all the materials to be used. For example, the cylinders in automobile engines are generally lined with steel or a steel alloy; the pistons, which are often

---

In the Celsius scale (formerly called the centigrade scale), the ice point is zero and the steam point is 100; each 1/100th between these two points equals 1 Celsius degree (1°C).

The increments by which substances expand when heated from ice point to steam point are different for different substances and are called coefficients of linear expansion.

---

made of aluminum, are designed to fit precisely to a very small tolerance, and when doing so the greater expansion coefficient of aluminum (0.000024) must be taken into account. Similarly, long steel bridges are provided with rockers or rollers at the ends so that the spans can expand and contract, and expansion joints are provided between the concrete slabs of a road.

Suppose you were engineering a 1000-foot steel bridge. How much expansion would you need to allow for between a winter temperature of −10°C and a summer temperature of +40°C?

The temperature spread you must consider is 50°C. Your engineering reference or computer program tells you that the expansion coefficient of steel is 0.00001.1/°C, so your total expansion (and contraction) would be 0.000011 × 50; multiply by 1000 ft (the total length of the bridge), and you find that your bridge will expand and contract with temperature by a total of 0.55 ft.

When we talked about mercury thermometers, we saw that mercury (a liquid) expands as temperature increases. Most liquids behave this way, but water is an exception. Between the ice point and about +4°C, water *contracts* very slightly. With further increases in temperature, it expands. This property, together with the

fact that it freezes at a moderate temperature, makes water unsuitable for use in a thermometer.

Exploration 6.2

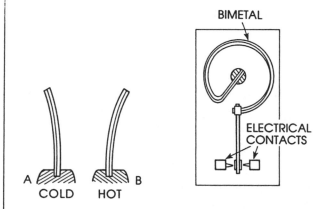

The very slight increase in length of a moderate-sized object may be magnified by using a device called a *bimetal*. It consists of a strip of iron and a strip of brass welded or riveted together along their length. The difference in their amounts of expansion shows up easily because the double strip bends into a curve when its temperature changes. The movement may be used to turn a pointer, to regulate a valve or to close a switch. A bimetal forms the main element of a *thermostat*.

When water is heated to just beyond the ice point, it contracts slightly, then expands with further increases in temperature.

The fact that water reaches its maximum density at a point just above its freezing point has important consequences for aquatic plants and animals. In winter, the water at the surface of a pond is in contact with the cold air. On cooling, it becomes denser and sinks. The water circulates in this way until all of it is at 4°C; only then do the top layers get colder and solidify (freeze). This means that fish and other aquatic life have a better chance to survive, since it lengthens the time before the body of water freezes solid from top to bottom.

Expansion and Contraction of Gases

Mercury is the most convenient substance to use in a thermometer, but a gas thermometer gives a better idea of what happens to molecules when they are heated or cooled. In one kind of gas thermometer, a pressure gauge attached to an enclosed vessel containing gas shows the changes in pressure that occur as the vessel is cooled. In another type of gas thermometer, a piston is used to keep pressure constant within the vessel (Figure 6.1), and changes in the volume of the gas are measured as gas is cooled.

Experiments with the first type of gas thermometer and different kinds of gases have shown that starting from the ice point on the Celsius scale (0°C), pressure exerted by a gas is reduced by $1/273$ for each degree of cooling. With the other type of gas thermometer, it was shown that the volume of any kind of gas is reduced by $1/273$ for each degree of cooling. In other

Figure 6.1

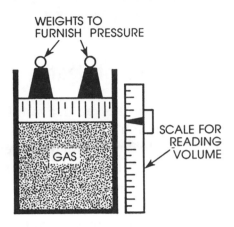

WEIGHTS TO FURNISH PRESSURE

GAS

SCALE FOR READING VOLUME

words, at -273°C, a gas would exert no pressure and take up no space except the negligible space taken up by the molecules themselves.

This is the basis on which -273°C was determined to be absolute zero, and it is the basis for the Kelvin scale of temperature measurement in which 0 is taken to be 273°C below the ice point.

The kinetic theory of gases explains what is happening; this theory is more fully explained in Chapter 8, Pressure. Briefly, a gas in a container exerts pressure because of the kinetic energy of its moving molecules. When the molecules strike the sides of the container, they push on it (create pressure). The lower the temperature of the gas, the slower the movement of the molecules, that is, the less their kinetic energy. If all the kinetic energy could be taken away by bringing the molecules to a standstill, they would cease to exert any pressure. In the laboratory, physicists have been able to come within a few 10 millionths of absolute zero.

The pressure exerted by any gas and the volume of the gas are reduced by $1/273$ for every degree Celsius the gas is cooled; this is the basis for establishing absolute zero as 273°C below ice point.

Conduction of Heat

Heat is transferred in a number of ways: by conduction, by convection, and by radiation. Regardless of the method, only heat is transferred. Cold is merely the absence of heat, so there can be no such thing as a transfer of cold.

Heat always moves from the warmer to colder. We already said that heat is a form of energy, the kinetic energy of the random motion of molecules. When two objects of different temperatures are put in contact, the faster-moving molecules of the warmer one collide with the slower-moving molecules of the colder one, transferring some of their motion to it. The warmer object loses energy (drops in temperature) while the cooler one gains energy (rises in temperature). When the two objects or substances are equal in temperature, the transfer process stops.

One way that heat passes from one place to another is by the handing-on of molecular motion through a substance; this is called heat **conduction.** Not all materials conduct heat at the same speed. Metals are called good conductors because they transfer heat fast. Stone is a moderately good conductor, while wood, paper, cloth, and air are poor conductors. Different materials are assigned numbers that give their relative rates of conduction com-

pared to silver, which has arbitrarily been made the standard and is given a **coefficient of heat conduction** of 100. Heat conduction coefficients of some common materials are given in Table 6.1. Note that the conduction coefficient of a vacuum (a space from which all air has been extracted) is zero.

Exploration 6.3

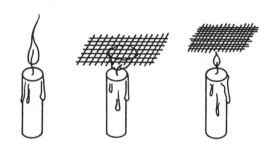

Bring a piece of wire screen down onto a candle flame. The flame will be cut off above the screen, since the heat is conducted away before the vaporized wax there can be ignited. Early in the nineteenth century, there were many disastrous explosions in British coal mines, caused by ignition of mine gases by the open flames of the miners' lamps. The great chemist, Sir Humphry Davy, suggested surrounding each lamp with a fine wire screen. It worked.

Only heat can be transferred; cold, being the absence of heat, cannot be transferred.

The ability of materials to conduct heat is expressed by their coefficient of heat conduction, which is based on an arbitrary selection of silver as the standard, with a conduction coefficient of 100.

Table 6.1   Heat Conduction Coefficients of Selected Materials

Material	Coefficient
silver	**100** (standard)
copper	92
aluminum	50
iron	11
glass	0.20
water	0.12
wood	0.03
air	0.006
perfect vacuum	0

EVERYDAY PHYSICS: 6

If you're out skating on a cold winter day, a cup of hot chocolate makes a welcome hand warmer. You'll probably curl both of your hands around it, feeling the warmth seeping through to you. You'll think of this as warming your hands.

In summer, you may hold a chilled soda can to your cheek in much the same way. And you may think of the can as cooling your cheek. But remember, cold is only the absence of heat; it cannot be radiated, conducted, or otherwise transferred. What you are actually doing is warming the can with the heat from your cheek. What you feel as a gain in cold is actually a loss of body heat.

A material that conducts heat poorly may be used to insulate, that is, to greatly reduce the transfer of heat by conduction or convection. Many materials used to insulate actually do so by means of trapped air.

You are making use of different rates of conduction when you wrap a cloth pot holder around the handle of a metal frying pan. Manufacturers of frying pans can make use of the conductive properties of different materials when they encase metal pot handles in wood or some other poorly conductive material.

Materials that conduct heat poorly are often called *insulators.* Air, which has a conduction coefficient of only 0.006, makes an excellent insulation material when it is trapped within spaces. Woolen clothes, furs, synthetic foams, and materials filled with loose fibers, feathers, or down insulate by means of the air trapped within them. When you layer your clothes on a cold day, it is the air between layers that is keeping you warm.

A vacuum bottle (Figure 6.2) makes use of the perfectly nonconductive properties of a vacuum. Such a bottle contains a double-walled inner liner that was pumped clear of air and then sealed. The walls of the liner are silvered to "reflect" the heat. The same principle will keep the contents either hot or cold.

Convection

Most liquids and all gases are poor conductors. However, they can transfer heat in another way: by **convection,** the mass movement of a heated liquid or gas. A gas (for example, air) that is heated expands, becoming less dense. Above a bonfire, for example, the air becomes warmed, expands until it is lighter than the surrounding air, and rises. Cool air flows in from all sides to take its place. Soon a continuous circulation is set up.

When you set a pot of water on a stove, the water nearest the flame heats up, rises, and is replaced by cooler water until eventually all the water is hot. In a home heating system, air sinks or drops down as it cools, enters the furnace jacket, is heated, and rises through the ducts or radiators, cools, drops again, is reheated, and so on.

Winds are the result of rapid convection currents in the atmosphere. Near the equator, the intense heat of the sun causes a general rising of the warmed air, while cooler air flows toward the equator to replace it. Many other factors modify this effect, including ocean currents such as the Gulf Stream, which are also produced by convection.

Figure 6.2

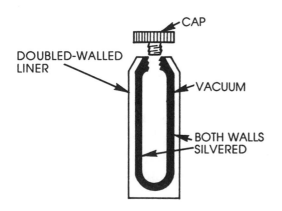

Convection is a manner of heat transfer by means of the mass movement of heated liquids and gases.

Exploration 6.4

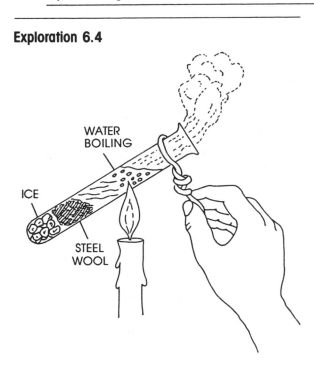

WATER BOILING

ICE

STEEL WOOL

Hold some small pieces of ice in place at the bottom of a vial or test tube containing water by pushing in a tuft of steel wool on top of them. Heat the water near the top of the tube with a candle or a gas flame. Soon the water will start to boil, yet the ice will not melt. The warmed water is already at the top, so no convection takes place, and the conduction by water is very small; altogether, then, very little heat is transferred to the ice.

Thermal Radiation

When you sit by an open fireplace, you feel warm. Yet heat is not reaching you by convection; in fact, the flow of air is in the opposite direction, as cool air from the room flows into the fireplace and heat goes up the chimney. The conductive value of air is negligible. So how do you feel warmth? What is happening is that the fire is sending out waves much like light waves. You cannot see these waves, but they are traveling through space by a process called **radiation.** Unlike conduction and convection, radiation does not require direct contact between substances.

Almost all the energy available to us on earth comes to us by radiation from the sun. When we burn coal or fuels, we are simply releasing potential chemical energy from the sun that was stored in plants millions of years ago. A hydroelectric plant in which water turns the turbines that generate electricity derives its energy from the great cycle of evaporation maintained by the sun: Water is lifted from lakes and oceans, then condensed to rain that feeds streams and waterfalls.

The sun is an extremely hot body, and so it radiates enormous amounts of heat. However, all things—stones, buildings, and your body, too—radiate heat to their surroundings. The air in a crowded theater becomes warm because each person in it is radiating heat equal to that given out by a 75-watt lightbulb. Even a cake of ice sends out radiant energy. Your hand feels cool near the ice because, being at a lower temperature, it sends less energy back to your hand than it gets from it.

The amount of energy an object can radiate depends not only on its temperature but on the nature of its surface. Dark, rough surfaces send out more radiation

---

When you burn wood in your fireplace, you are releasing heat energy radiated by the sun and stored in the wood; the process by which you feel warmth is called radiation.

---

than smooth, light-colored ones at the same temperature. The reverse is also true: dark, rough surfaces absorb radiation better than light ones. In the tropical sun, a dark-colored jacket feels warmer than a light-colored one of the same weight. In sunshine, dirty snow melts faster than clean snow.

The so-called **greenhouse effect** is a result of the ways in which different materials absorb and transmit different kinds of light and heat waves. The kind of greenhouse used in raising plants has a glass roof. The rays of the sun pass readily through the glass, warming the soil inside. The soil, being warmed, also emits rays, but these longer rays cannot get through the glass. The greenhouse acts as a heat trap.

The earth's atmosphere works like the glass in the greenhouse. Visible short-wave radiation from the sun passes through the atmosphere to warm the earth. The earth sends back warm-wave radiation that cannot completely penetrate the atmosphere, particularly the carbon dioxide in it. Some of the heat is returned to the earth's surface. The higher the amount of carbon dioxide in the atmosphere, the more heat is retained. The burning of fossil fuels raises the carbon dioxide level of the atmosphere so that more heat is retained. Even a small rise in average global temperature affects the level of the seas and other aspects of weather and climate. Scientists believe that this greenhouse effect will raise average global temperatures by as much as 9°F by the year 2000.

Exploration 6.5

Cut out the ends of a tin can and paint one with flat black paint. Set both disks out in the sun and, after several minutes, feel each one and notice how much hotter the blackened one is, because of its greater rate of heat absorption.

Ordinary glass is transparent to visible light, but not to the longer, invisible waves given off, say, by soil that has been warmed by the sun. This fact is made use of in a **greenhouse** or hot frame. The sun's rays pass readily through the glass roof and are absorbed in the soil within. This, being warmed, then emits rays of its own. But these are mainly long waves that cannot get out through the glass, and so the greenhouse acts like a heat trap. In localities where there is enough sunshine in winter, houses can be heated by the same principle.

All objects radiate heat, including your body; the hotter the object, the more heat it radiates.

Testing Your Knowledge

6.1 Which is larger, a Fahrenheit degree or a Celsius degree? What is the ratio of their sizes?

6.2 The normal temperature of the human body is taken to be 98.6°F. How much is this on the Celsius scale? On the absolute scale?

6.3 Examine the Table below and explain why a Pyrex dish can be taken directly from the oven and plunged into cold water without cracking.

Table 6.2   Coefficients of Linear Expansion for Selected Solids

aluminum	0.000024
brass	0.000019
iron or steel	0.000011
ordinary glass	0.000009
Pyrex glass	0.000004
Invar (an alloy)	0.0000009

6.4 Why is an automobile engine noisy until it has warmed up to running temperature?

6.5 When a piece of iron with a cavity inside it is heated, does the hole become larger or smaller? Explain.

6.6 An aluminum piston in a car engine is 2³/₄ inches in diameter. How much does its diameter increase when warmed from 10°C to its normal operating temperature of 170°C?

6.7 A silver spoon and a book are both at room temperature. The spoon feels colder to the touch because (choose one)
   a. it is made of a denser material.
   b. silver is a very good heat conductor.
   c. silver is an almost pure material.
   d. the book has the greater weight.

6.8 Ice is placed in the *upper* part of an ice chest because
   a. it is easier to reach there.
   b. the water formed in melting can run out more readily.
   c. convection will distribute the cooled air.
   d. it will come in direct contact with the food.

6.9 Every object at a temperature above absolute zero
 a. must receive heat by convection.
 b. radiates energy.
 c. occupies less space than it would at absolute zero.
 d. is a good heat insulator.

6.10 The best absorber of radiation is a body whose surface is
 a. glossy and gray.
 b. white and fuzzy, like wool.
 c. a mirror.
 d. dull black.

6.11 We know that the energy we receive from the sun is not transported by conduction or convection because
 a. interplanetary space is a good vacuum.
 b. air is less dense at high altitudes.
 c. the sun is gradually cooling off.
 d. there are always some clouds in the atmosphere.

CHAPTER 7

Thermal Energy

KEY TERMS FOR THIS CHAPTER

thermal (heat) energy
specific heat
calorie
kilocalorie
British thermal unit (Btu)

heat of fusion
evaporation
heat of vaporization
heat–work equivalent
conservation of energy principle

When people thought that heat was a fluid that is transferred from warm objects to cold ones, they could not account for some ways in which heat was produced—for example, by friction. Once scientists defined heat as a form of energy, they could explain how mechanical energy could be transformed into heat. The opposite process, turning heat into mechanical work, is the basis for the steam, diesel, gasoline, and other heat engines that our present-day industrial world relies on. This chapter explains how heat is energy and how that energy can be applied.

Measuring Thermal Energy (Quantity of Heat)

In Chapter 6 you learned how temperature may be measured, but that is not the whole story of measuring heat. Temperature tells us what degree of heat is present in something but does not reveal how much heat energy something contains. A cup of boiling water has a higher temperature than a tub full of warm water, but the tubful contains more heat energy. For instance, if you put a large block of ice into a tub of warm water, it would eventually melt completely, but pouring a cupful of boiling water over a block of ice would melt only a small portion of the block.

Heat energy is also called **thermal energy.** The amount of thermal energy transferred to or from a substance when its temperature is changed by a certain amount is called quantity of heat. It depends on the nature of the material. Suppose you could take an iron ball and a lead ball of the same size, heat them both to the temperature of boiling water, and lay them on a block of wax. Even though lead is denser than iron—that is, has a greater mass—you would find that the iron ball melts a considerable amount of wax while the lead ball melts hardly any. Apparently, different materials give up different amounts of heat while cooling through the same temperature range.

Experiments have shown that the quantity of thermal energy given up or acquired when a body changes its temperature is proportional to the mass of the object, the amount by which its temperature changes, and its **specific heat.** Specific heat is like the coefficients of conduction and linear expansion in being a fractional number assigned to a substance on the basis of its comparison to an arbitrary standard. For specific heat, water is the standard and is assigned a value of 1. Table 7.1 gives specific heats of common materials.

Expressed in the terms of a formula: Q (thermal energy) = s (specific heat) $\times$

Table 7.1   Specific Heats of Common Materials

Water	1.00 (standard)
alcohol	0.65
aluminum	0.22
glass	0.20
iron	0.11
copper, brass	0.09
silver	0.06
lead	0.03

m (mass) $\times$ t (temperature change), or simply: $Q = smt$

From this equation you can infer that Q is a measure of total energy, while t (temperature) is really the average energy per molecule. If the equation is rearranged to solve for t, the result is the quantity of heat per unit of mass. No special unit such as degrees need be attached to s, which is merely a ratio.

The fact that the specific heat of water is considerably greater than that of most other substances makes water act as a sort of storehouse for heat. For example, this property of water explains why daily and seasonal temperature changes are less severe on islands and along seacoasts. When the sun is strong, the ocean takes up a great deal of heat without changing its temperature appreciably. When the sun is weak, or at night, the stored heat is given up to the surroundings, preventing the temperature from going as low as it otherwise might. (Watch regional high/low weather maps on TV and observe the

---

The amount (quantity) of heat energy an object gives up or acquires when its temperature changes depends on its mass, the amount of temperature change, and the specific heat of the substance(s) of which it is composed.

The comparatively high specific heat of water enables large bodies of water to serve as equalizers of temperature.

---

range of temperature changes in areas near large bodies of water as compared with regions farther inland.)

Heat Units

Heat quantity (*Q*) is the nature of thermal energy, but it is convenient to have a special unit for it. In the metric system, the basic unit of thermal energy is the **calorie,** which is defined as the amount of heat entering or leaving one gram of water when its temperature changes by 1°C. When dealing with large quantities of heat, scientists use the **kilocalorie** (kcal); one kilocalorie equals 1000 calories.

Do not confuse the calorie and kilocalorie used in physics with the kilocalorie used in specifying food values (usually, to confuse things further, called simply a calorie). When dietitians say a slice of bread contains 80 [kilo]calories, they mean that in the process of digestion it furnishes that amount of heat energy. The average adult male burns about 14 calories per 24 hours per pound of body weight in sedentary and maintenance activity; the average female uses about 11. The amount of calories "burned" increases with muscular exertion.

To get back to physics, in the English system, heat energy is measured in **British thermal units (Btu);** 1 Btu is defined as the amount of heat needed to change the temperature of a pound of water by 1°F.

One Btu is equivalent to 252 calories or about 0.25 kilocalories.

Exploration 7.1

If we put hot and cold substances in contact and prevent heat transfer with the surroundings, everything will finally come to a common temperature. Provided that no conversion of heat to other forms of energy (or vice versa) occurs, it will be true that *the total heat given up by hot bodies will equal the total heat taken on by cold bodies.* If we know all the other circumstances, we can use this statement to predict the final temperature.

Try this experiment. Heat a 200-gram brass ball to 80°C, then plunge it into 150 grams of water at 20°C. What will be the final equilibrium temperature of both?

Call the final temperature *t*°C. It must lie somewhere between 20° and 80°C. Using $Q = smt$, the heat given up by the brass in cooling from 80° to $t°$ will be (use Table 7.1 for finding s): $0.09 \times 200 \times (80 - t)$. Similarly, the quantity of heat taken on by the water will be $1 \times 150 \times (t - 20)$. Setting these two heat quantities equal, we can solve for t: $0.09 \times 200 \times 80 - 0.09 \times 200 \times t = 1 \times 150 \times t - 1 \times 150 \times 20$; finally, $t = 26.4°C$.

The calorie used in physics is defined as the amount of heat entering or leaving 1 gram of water when its temperature changes by 1°C; it is not the same as the calorie used in nutrition.

Heat of Fusion

A large change in temperature may also change the physical state of a material. For crystalline substances, the temperature at which the solid form melts is the same as that at which the liquid form freezes when cooled. We call this the ice point. (Water is a crystalline substance, although you may not realize it unless you think of snowflakes.) Noncrystalline substances such as wax or glass have no definite point of melting or solidifying; for example, butter gradually softens as its temperature is raised and gradually hardens as its temperature is lowered.

Thermal energy is required to melt ice. Putting ice cubes into your cola cools it because energy is taken from the cola to melt the ice. If you put ice into water at room temperature, the temperature of all the water in the container will finally come to ice point (0°C or 32°F) so long as any ice remains. Putting in more ice or more water will not alter the overall temperature; that will begin to rise only when all the ice has melted (see Figure 7.1).

The quantity of heat (thermal energy) needed to melt one gram of a given substance without producing any change in temperature remains constant and is called the **heat of fusion** of that substance. For water, the heat of fusion is about 80 calories per gram or 144 Btu per pound. Because energy cannot be destroyed but only transferred (the principle of the conservation of energy), this same amount of

Figure 7.1

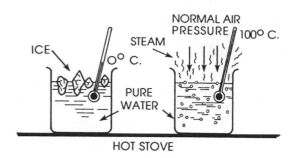

heat must be given off when one gram of material solidifies.

When a gallon of water freezes, it *gives up* as much heat as would be produced by burning an ounce and a half of good coal. In other words, the formation of snow and ice in winter actually warms the air to some degree. This principle has been used to keep vegetables in storage from freezing: Large tubs of water are placed in the storage area. The water, being pure, freezes at a higher temperature than the fluids in the vegetables, and in the process of freezing it may give off enough heat to prevent the temperature in the storage area from going lower.

Crystalline materials generally increase sharply in volume when the liquid freezes. That is why ice floats; its volume is greater than that of water. That is also why water pipes or auto radiators burst when the water in them freezes, and why rocks and pavement are split by freezing water.

Once a liquid has been brought to boiling point by heating, additional heat will not make the liquid hotter but will increase the rate of vaporization instead.

Heat of Vaporization

Evaporation of a liquid consists of the escape of molecules from its surface. Only the faster, more energized molecules can get away from the attraction of the others. When they do so, the average speed of the molecules in the liquid is lowered and the temperature of the liquid falls. Thus evaporation produces a cooling effect.

Evaporation at the free surface of a liquid occurs at all temperatures. (You may have seen a lake or pond "steaming" on a chilly fall morning when the air temperature has dropped sharply overnight.) When the pressure of the vapor equals the pressure of the surrounding air, bubbles of vapor form *all through* a liquid. We call this state *boiling*. If the liquid was brought to a boil by being heated, continuing to apply heat will not make it hotter but will merely make it boil away faster (that is, will make molecules escape faster). At this point, the energy supplied by heating the liquid is used by the molecules in separating from each other.

A given amount of heat, which is different for different substances, is carried away for each gram of liquid that vaporizes. This amount of heat is called the **heat of vaporization.** An equal amount of heat is *given off* whenever a gram of vapor condenses (turns to liquid upon being cooled) at the steam point. The heat of vaporization/condensation for a material is calculated at sea level. For water at its sea-level steam point (100°C, 212°F), the heat of vaporization/condensation is 540 cal/g (calories per gram) or 970 Btu/lb. (The value for vaporization/condensation occurring at levels other than the steam point is slightly different.)

At sea level, the air pressure acting on the surface of water is 14.7 lb/in² and boiling begins at 100°C (212°F). At higher temperatures, boiling begins at a lower temperature (that is, vapor pressure and surrounding air pressure equalize sooner). At the top of Pike's Peak (14,000 feet above sea level), for example, boiling begins at only 85°C. Cooking food in an open vessel becomes difficult under such conditions. So does calculating cooking time. A closed vessel will allow the pressure of the vapor to build up inside, raising the steam point. (This is the principle of the pressure cooker; raising the pressure on the liquid by sealing it in the cooker raises the steam point so the contents cook faster.) Condensation is slightly different at other temperatures than the steam point.

The great play of evaporation and condensation of water in the atmosphere is one of the most important factors affecting the weather. When most air is cooled, the vapor may condense into a fog of tiny, slowly settling droplets. Fog forming at a certain distance above the ground becomes clouds, and when the droplets are large enough, they fall as rain. In winter, moisture in the air may go directly into its solid state, settling on chilled surfaces as frost. If the moisture solidifies in the air itself, we have individual crystals that fall as snow. Hail consists of raindrops that have frozen as they passed through cold layers of air on their way to the ground.

If you quickly take the cap off a cold bottle of soda, fog will form in the neck of the bottle as the warm air from the room meets the soda. In cloud chambers used in scientific work, fog is intentionally produced, for example, to show up the paths of atomic particles.

EVERYDAY PHYSICS: 7

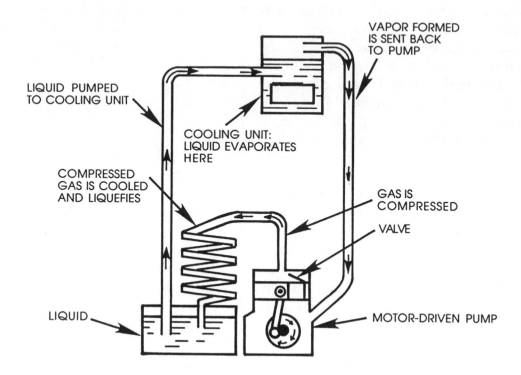

VAPOR FORMED
IS SENT BACK
TO PUMP

LIQUID PUMPED
TO COOLING UNIT

COOLING UNIT:
LIQUID EVAPORATES
HERE

COMPRESSED
GAS IS COOLED
AND LIQUEFIES

GAS IS
COMPRESSED

VALVE

LIQUID

MOTOR-DRIVEN PUMP

A heat pump is a refrigeration device that can either heat or cool a house. A cold refrigerant circulates through pipes that are exposed to the outside air. To heat the house, it absorbs heat from the air and transfers it to the air inside the house. To cool the house, the flow of the refrigerant is reversed so that it draws heat from the inside air and transfers it to the outside air, which is cooler than the refrigerant.

A refrigerator is really a heat engine in reverse; that is, mechanical work is made to produce a difference in temperature. A motor-driven pump compresses an easily liquefied gas such as methyl chloride or Freon. The compressed gas is passed through coils that are cooled by air (or sometimes by water). In cooling, the gas becomes a liquid that is pumped into coils or channels at the back of the food storage compartment or in the walls of the freezer compartment. Within these coils, the pressure is reduced to make the liquid vaporize. The vaporization process takes heat from the surroundings, cooling the air inside the compartment as well as the food stored in it. The vaporized fluid returns to the pump and the cycle is repeated. In a gas refrigerator, a small gas flame performs the function of the pump.

Exploration 7.2

Put some water on a large cork. On top of the cork, set a piece of aluminum foil shaped into a small dish. Pour a little alcohol or ether into the dish (keep away from open flames) and make it evaporate rapidly by vigorous fanning. Enough heat will be carried away to turn the water into snow, or even to freeze the dish firmly to the cork.

Heat–Work Equivalent

In any mechanical process, a certain amount of energy is wasted (dissipated) in the form of heat. The bearings of a machine become warm, a pump for compressing air is hotter than friction alone can account for, a nail is warmed when a hammer strikes it. The mechanical energy that is lost appears again in the form of heat energy. Experiments show that whenever a given amount of mechanical energy disappears, a fixed quantity of heat appears in its place, regardless of whether the change is brought about by friction (rubbing of materials against one another), compression of a gas, stirring of liquids, or some other factor. This fixed rate of exchange, or **heat–work equivalent** has an experimental value of 4.18 joules per calorie or 778 foot-pounds per Btu.

Just as work may be transformed into heat, heat may be transformed into work. Heat engines such as steam engines, turbines, and jet engines operate by this principle. When heat is transformed into work, the heat–work equivalent is the same as that for the reverse: 4.18 joules per calorie or 770 foot-pounds per Btu. Through this value and through knowing the specific heat of different materials, it is possible to calculate how much work, say, the burning of a particular quantity of a particular fuel can accomplish.

The heat–work equivalent is a demonstration of the general **principle of conservation of energy** first defined by J.R. Mayer in the nineteenth century, namely, energy that disappears in one form must reappear in another. This principle applies not only to mechanical energy but to all forms—thermal (heat) energy, chemical energy, electrical energy, and so on.

Testing Your Knowledge

7.1 The sparks from the flint of a cigarette lighter are red hot, yet they do not burn the skin of your hand. Why?

7.2 How many Btu does it take to heat a 5-lb iron from 65° to 330°F?

7.3 How many Btu does it take to change 1 lb of ice at 32°F to steam at 212°F?

7.4 It may be said that (a) boiling is a cooling process and (b) freezing is a warming process. Explain.

7.5 Why is a burn by live steam at 100°C worse than one by boiling water at 100°C?

7.6 Can you cool a perfectly dry object by fanning it?

7.7 A glass of cold liquid is observed to "sweat" on a warm, moist day. Where does the water come from?

7.8 Wet clothes hung out on a line in winter are often observed to "freeze dry." Explain.

7.9 What becomes of the KE of an automobile when it is brought to a stop by the brakes?

7.10 How much heat energy, in Btu, is used by a 160-lb man when he climbs a mountain 3900 ft high?

7.11 In level flight, a certain rocket motor has a thrust of 3000-lb force. If a pound of rocket fuel can furnish 30,000 Btu when burned, how far will 1 lb of fuel drive the rocket, assuming that $1/4$ the energy of this fuel is changed to mechanical work?

7.12 Is the temperature of the steam under pressure in the boiler of a steam engine higher or lower than 100°C?

7.13 Does a refrigerator, operated with the door open, produce any net change in the temperature of the room? If so, in what direction?

CHAPTER 8

Pressure of Liquids and Gases

KEY TERMS FOR THIS CHAPTER

pressure

hydraulics

vacuum

barometric pressure

Boyle's law

buoyancy

flotation

displacement

resistance

specific gravity

streamlining

Bernoulli's law

Earlier, we talked about force as being a push or a pull. In this chapter, we discuss force as it relates to liquids and gases. Many devices and machines you use every day make use of physical principles that apply to liquids and gases, including complex machines like airplanes and automobiles and simple devices like the push-button spray bottle you use when cleaning windows.

What Is Pressure?

In everyday speech, we often use "force" and "pressure" interchangeably. In phys-ics, these words have more exact meanings. You have already learned how the amount of force used to move matter is measured in terms of work. Objects also exert force (push/pull) when they are lying still. If you put a 10-lb weight on a table, it exerts a downward force of 10 lb on the tabletop. Downward (gravitational) force can be measured in weight units, and simple arrangements such as springs, pulleys, and weights allow us to measure force in other directions.

The **pressure** that an object exerts depends on the area of contact. It is measured by dividing the force by the area of

---

The amount of pressure exerted on a surface by an object at rest depends on the area of contact.

---

the surface on which it acts. If a 10-lb weight that has a bottom area of 5 square inches (5 in²) contacts a tabletop evenly all over its 5-in² face, the pressure between it and the table will be:

$$\frac{10 \text{ lb}}{5 \text{ in}^2} \text{ (divided by)}$$

$$= 2 \text{ lb/in}^2 \text{ (pounds per square inch)}$$

Suppose the weight is standing on its smaller face, which has an area of only 2.5 in². The pressure then would be 10 lb/2.5 in² or 4 lb/in²—twice as much as before. In this case the same force is spread over only half the area.

In general, then, we can say that:

$$p \text{ (pressure)} = \frac{F}{A} \frac{\text{(force)}}{\text{(divided by)}} \quad \text{(area)}$$

Pressure is a derived quantity. It can be measured and expressed in any combination of an area unit (itself derived from a length unit) and a weight unit belonging to the same system: lb/ft², kg/cm², and so on.

Liquid (Hydraulic) Pressure

So far we have been talking about a solid object. A liquid pushes on the sides as well as on the bottom of its container. That's why barrels and water tanks have

to be reinforced. But it's also true that a liquid at rest presses upward on anything placed in it. You can demonstrate this for yourself: Push the closed end of a water glass or the flat side of a plate against the surface of the water in a dishpan, and you will feel the upward thrust of the water against it.

At any point within a liquid at rest, the pressure is the same in all directions: up, down, and sidewise. However, pressure in all directions increases with depth. Again, you can demonstrate this for yourself (Exploration 8.1).

Exploration 8.1

Push the closed end of a tumbler or empty tin can beneath the surface of water in a bowl and you will actually feel the upward thrust of the water on the bottom.

Suppose you were pouring water into a tall, narrow container that has a bottom 1 in² in area. If you pour 1 lb of water into it (about a pint), the force on the bottom will be 1 lb/in². Now pour in a second pint of water. The bottom is now supporting 2 lb of liquid, so that pressure on it is 2 lb/in². Reasoning from this very simple example tells us that the pressure at any point in a liquid that has a free (open) surface is directly proportional to the depth below the surface.

Liquids exert force on the sides as well as the bottom of anything they are in.

In a liquid at rest (a still liquid), the amount of pressure exerted at any depth is equal in all directions.

When we talk about depth in this way, we are talking about distance *straight down* from the level of the free surface of the liquid to the level of the place in question. Even if the vessel or pipe slants, this is the way the depth must be measured. In the teapot shown in Figure 8.1, the free surfaces of the liquid in the pot and the liquid in the spout stand at the same level, even though the weight of water is greater in the pot than in the spout. Pressure depends only on the vertical depth and not on the size or shape of whatever contains the water. No water flows one way or another at the place where the pot and spout join, so the pressure must be the same from both sides and so must the depth.

Figure 8.1

The principle that water seeks its own level is utilized in municipal water supplies. If the lake or reservoir from which the water comes is higher than any building in the town (for example, if it is on a mountain), water will flow into the mains and buildings on this principle. Or, water can be pumped into a standpipe (water tower) so that the height of the water in the tower provides pressure that makes the water flow to the buildings it supplies. Buildings higher than the standpipe level must have an auxiliary pump to supply water to the upper floors.

Computing Hydraulic Pressure

To compute amount of pressure at any depth in any liquid, we need to consider the density of the liquid: Pressure is proportional to both depth and weight. Since weight is proportional to density, doubling the density would double the weight of any column of liquid. Therefore,

p (pressure) = h (height, or depth) $\times$ D (density)

or

$$p = hD$$

The unit used for D must correspond to the unit used for h; that is, if h is given in feet, D must be in pounds per cubic foot (lb/ft^3); if h is given in centimeters, D must be in grams per cubic centimeter (g/cm^3), and so on.

Densities of some common liquids are:

	lb/ft^3	g/cm^3
gasoline	44	0.70
water	62.4	1.00
seawater	64	1.03
mercury	850	13.6

At any point in a liquid that has a free (open) surface, pressure is directly proportional to the depth below the surface measured as a straight vertical line.

Exploration 8.2

You are standing on a dam watching the water spilling over. What is the pressure on the side of the dam at a point 20 ft vertically below the water surface? In the formula $p = hD$ put $h = 20$ ft and $D = 62.4$ lb/ft³, getting $p = 20 \times 62.4 = 1248$ lb/ft². Notice that since h was given in feet, we had to use the density in corresponding units, that is, in pounds per cubic *foot*. The result is then in pounds per square foot. Now that we have the answer, we are at liberty to change it to any other units we like. Very often, pressure in the English system is given in pounds per square *inch*. Since there are 144 square inches in a square foot, we can change our result to these units by dividing by 144. Then we have $p = 1248/144 = 8.67$ lb/in².

Exploration 8.3

What is the *total force* on the bottom of a swimming pool 80 ft long and 25 ft wide, filled to a depth of 5 ft? What is the force on one of the sides?

The total force is the pressure (force per unit area) multiplied by the area on which it acts. Then $F = hDA$, or $F = 5 \times 62.4 \times 80 \times 25 = 624,000$ lb, or 312 tons. The pressure on a side will vary from zero at the surface to its greatest value at the bottom. To get the total force on a side, we must then use the *average* pressure, or the pressure *halfway down*. In this case, we must take $h = 2.5$ ft. Then $F = 2.5 \times 62.4 \times 80 \times 5 = 62,400$ lb $= 31.2$ tons.

The fact that hydraulic (fluid) pressure is exerted evenly in all directions is the principle underlying many everyday applications. In any kind of hydraulic press, pressure is applied mechanically to a small piston, and this same amount of pressure then acts on every part of the inside surface of the system, including the large piston that does the pressing. If the area of the large piston is, say, 100 times that of the smaller one, the total force on the large one will be 100 times whatever force is applied to the small piston (Figure 8.2).

Figure 8.2

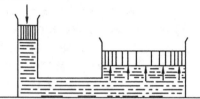

Hydraulic presses are used in making bricks, glassware, and some metal parts. Large machines of this kind may be capable of exerting forces of 10,000 tons or more. The car lift used at your service station is a kind of hydraulic press; in this case the pressure is used to lift something. When your hair stylist pumps up the barber chair with one foot, it is hydraulic pressure that makes the chair rise—and you with it.

Pressure of Gases

Recall from Chapter 2 that there are three states of matter: solids, liquids, and gases. We have talked about pressure as it relates to solids and liquids; gases also exert pressure. As we used water in discussing fluid pressure, we will use air as the major example in talking about gases.

Air does have mass, although we aren't generally aware of it unless the air is in rapid motion. (When you are buffeted by a high wind, it is the air's mass that is pushing you.) We can weigh the mass of air by weighing a closed container, then pumping out the air and weighing the container again. If you took an empty 2-liter soda bottle and weighed it, then pumped out the air and weighed the bottle again, you would find a 2-gram difference. That is, air weighs about 1 gram per liter.

Since air has weight, it exerts pressure on anything immersed in it—including your body. You don't feel this pressure because it is counterbalanced by the overall pressure of air in your body cavities and tissues. At sea level, air pressure amounts to about 14.7 lb/in² (1031 g/cm²). This is over a ton per square foot.

Exploration 8.4

You can demonstrate the existence of air pressure by removing the air from one side of an exposed surface. Get a can (*not* a glass) that has a tight-fitting cover or an opening provided with a screw cap. Put a little water in the can, stand it in a pan of water and boil it vigorously, with the cover removed, so that the escaping steam will drive out the air. Weight the can down if it tends to upset. While the water is boiling, close the cap tightly, quickly transfer the can to a sink and run cold water over it to condense the steam inside.

Outside air pressure will crush the vessel in a spectacular way.

The condensing (turning to liquid) of some of the steam in the last experiment left a partial **vacuum** inside the can. A vacuum is simply a place not occupied by matter, or an empty space. For a long time, people believed that a vacuum had the mysterious power of "sucking" things into it. But how does the vacuum you create when you sip a soda succeed in getting a grip on the liquid in order to pull it up into your mouth?

Barometric Pressure

The condensing of steam within a container leaves a partial vacuum. A **vacuum** is a space with no matter in it. If the seal of a vacuum container is opened, air will rush into it to equalize the pressure within and without. This "sucking power" of a vacuum is the basis on which many pumps operate, and people were aware of it for a long time before a friend and pupil of Galileo demonstrated the underlying principle.

Back in the seventeenth century, it was found that no pump was able to raise water more than 34 feet above a well (that is, that a vacuum could "suck" only a 34-foot column of water). Evangelista Torricelli conducted experiments to find out why this was so. First, he decided to use mercury rather than water; since mercury is 13.6 times as dense as water, a height of only 2.5 feet instead of 34 feet would

---

A vacuum is a space with no matter in it. If the seal of a vacuum container is broken, air will rush in to equalize the pressure; a person is "sucked out" of a moving airliner when the door opens by the force of the air rushing *out*.

"satisfy the vacuum." He filled a glass tube a yard long with mercury, sealed one end, and held the other end closed with his thumb. Then he turned the tube over and set the open end of it in a large dish of mercury. When he took his thumb away, the mercury dropped away from the sealed end of the tube until its upper surface was about 30 inches above the surface of the liquid in the dish.

In descending from the top of the tube, the mercury left a vacuum behind it, and this vacuum was able to hold up a 30-inch column of mercury. Torricelli concluded that the mercury in the tube was not held up by any mysterious sucking action of the vacuum but by the *pressure of the outside air* on the mercury in the open dish.

Other people carried this experiment further. They found that when it was conducted on a mountaintop, the mercury in the tube did not reach so high a level, but when the same apparatus was returned to sea level, the mercury again reached 30 inches. These experiments not only explained how a vacuum works, they gave us a means of measuring air pressure: the **barometer.**

A more convenient form than the mercury barometer is the aneroid barometer. It uses a sealed metal vacuum chamber with a flexible cover. As changes in air pressure flex the cover, a lever system magnifies the motion, moving the pointer over a scale from which air pressure can be read directly (Figure 8.3).

The knowledge that the pressure of the air depends on altitude enables us to measure altitude by measuring air pressure. Altimeters used in airplanes are aneroid barometers with the scale marked directly in height units.

Figure 8.3

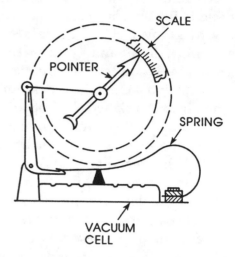

An even more familiar use of the barometer is in weather forecasting. Moist air is less dense than dry air, because water vapor is only about 5/8 as dense as dry air. Being less dense, moist air exerts less pressure, and so in moist weather the barometer falls. When the TV weather forecaster says the barometer is "steady," or "rising," or "falling," you can tell without hearing more that the weather will remain the same, clear up, or become rainy or snowy. When a storm is approaching, the barometer falls rapidly.

Gas Volume and Boyle's Law

Gases can be expanded and compressed. To force a gas into a smaller space, pressure must be applied. That is what happens, for example, when you pump air into a tire. The more pressure applied, the smaller the space the gas can be forced into. The Irish scientist Robert Boyle discovered the relationship now known as

Boyle's law: If the temperature of a gas is kept constant, the volume will be inversely proportional to the pressure. In other words, if the pressure is doubled, the volume is halved; if the pressure is tripled, the volume becomes one-third of what it was, and so on. Mathematically expressed, Boyle's law is:

$$\frac{V_1}{V_2} = \frac{P_2}{P_1}$$

These relationships are characteristic of *any* inverse proportion.

Exploration 8.5

Reading your car's manual, you find that your tire inflation pressure is 30 lb/in². The inside volume of the tire, assumed constant, is 0.95 ft³. What volume of outside air is needed to fill the tube to this pressure on a day when the barometric pressure is 15 lb/in²?

A tire gauge reads the pressure *above* atmospheric, so the total pressure on the air in the tire is 30 + 15, or 45 lb/in². Then, if V_1 is the volume that this amount of air occupies outside, we can make the proportion

$$\frac{V_1}{0.95} = \frac{45}{15}$$

Cross-multiplying, we get:

$$V_1 = \frac{0.95 \times 45}{15} = 2.85 \text{ ft}^3$$

Compressed air has many uses: keeping tires inflated, cushioning things from shock, operating air brakes and jack-hammers, and keeping water out of submarines.

Low pressures have their uses, too. The vacuum cleaner is a familiar example. Television picture tubes and computer CRTs have a vacuum inside. Modern pumps can reduce air pressure in a tube to less than one-billionth of normal air pressure and special methods can attain a pressure of one-billionth of one-billionth.

Buoyancy and Flotation

Earlier in this chapter you learned that a liquid exerts pressure equally on all sides, even pushing upward on the bottom of an object immersed in it. When an object is suspended in water, the pressures on opposite sides cancel each other out. Because pressure increases with depth, the upward force on the bottom of the object will be greater than the downward forces on the top. In other words, there is a net lifting force in operation; any object is lighter when in water than it would be if it were out in the air. The existence of such a lifting force is called **buoyancy.**

A large rock can be easily lifted from the bottom of a pond but becomes heavy the moment it clears the surface. Sitting in a well-filled bathtub, you can support your whole weight on your fingertips. The Greek philosopher Archimedes formulated the scientific law governing buoy-

---

Boyle's law: At constant temperature, the volume of a gas is inversely proportional to the pressure.

Archimedes' law of buoyancy: Any object immersed in liquid appears to lose an amount of weight equal to the weight of the liquid it displaces.

---

ancy nearly twenty-two centuries ago: Any object immersed in liquid appears to lose an amount of weight equal to the amount of liquid it displaces (pushes aside).

Exploration 8.6

Weigh an empty, corked bottle. Also weigh a pie tin. Put a pot in the pie tin and fill the pot brim full of water. Now lower the bottle carefully into the water, letting it float there. Remove the bottle, then the pot, and weigh the pie tin along with the water that overflowed into it. You will find the weight of water equal to the weight of the bottle, proving Archimedes' law for floating bodies.

It turns out that a body will float if its density is less than that of the liquid, otherwise it will sink. Recall what you learned about density, and you will understand why wood, ice, and gasoline can float on water, while iron, stone, and mercury sink.

For instance, a rock that has 0.5 cubic feet of volume will displace 0.5 ft³ of water. That amount of water weighs 0.5 × 62.4, or 31.2. Under water, then, the rock will weigh 31.2 lb less than it does when out of the water. For a body to float in water, its buoyant force must equal the whole weight of the object. It turns out that a body will float if its density is less than that of the liquid; otherwise, it will sink. Pine wood, gasoline, and ice all have

densities lower than water and will float; iron, stone, and mercury will sink. **Flotation** is the relative capacity of a material or object to float.

Seawater (salt water) has a greater density than fresh water and can float denser materials. You can illustrate this for yourself. A fresh egg will not float on water, because its overall density is greater than water's. If you put it into a glass of water, it will sink. But if you dissolve 2 tablespoonful of salt in the water, the egg will float, because the density of the water has been increased by the salt dissolved in it.

Once it was thought that iron ships would not float, since iron is "heavier than water." Actually, the *overall* density of a steel ship is less than that of water because the ship has a hollow interior. The total weight of a ship is called its **displacement.** For example, a ship that has a volume of 230,000 ft³ below the water line has a displacement of 230,000 × 62 = 14,260,000 lb, or 7,130 tons of salt water.

Sometimes the term **specific gravity** is used to indicate the density of material relative to water. In the metric system, the density of water is 1 g/cm³, which is numerically the same as its specific gravity. In the English system, the density must be divided by 652.4 to arrive at specific gravity.

The depth to which a floating body immerses itself in a liquid can be used as a measure of the density of the liquid. An instrument called a hydrometer is used for measuring the density of liquid solutions. It is a tall stick or tube that has a weighted end, which allows it to float upright, and a scale marked on its side. If you have watched a service station attendant testing the liquid in your radiator for its antifreeze level, you have seen a hydrometer in action.

Exploration 8.7

While you are beachcombing, you find a rectangular block of wood that measures 20 × 20 × 5 cm. When you float it flatwise, you see that 3 cm of the short side are under water. What is the density of the wood?

The block will sink until it just displaces its own weight of the liquid. The weight of water displaced will be 20 × 20 × 3, or 1200 gm, since water has a density of 1 gm/cm³. Then the density of wood will be this weight divided by the volume of the whole block, or 1200/20 × 20 × 5, which comes out equal to 0.6 g/cm³.

Buoyancy in Gases

The principle of buoyancy also applies to gases. A large hollow body like a balloon can displace more than its own weight of air, and so it can float in the air. However, air becomes less dense with its distance above sea level. A balloon filled with ordinary air at the same temperature as the outside air can float only to the height where the weight of the air becomes equal to its own weight. Filling balloons with gases that are lighter than air such as hydrogen or helium allows them to float higher.

Exploration 8.8

Make your own submarine: Get a tall jar with a flexible metal screw top and fill it with water. Fill a small glass vial about two-thirds with water, close the end with your thumb, and invert into the jar of water. Adjust the amount of water in the vial very carefully, drop by drop, until it just floats. At this stage the slightest downward push should send it to the bottom momentarily. Now fill the jar to the brim and screw the cap on tightly. When you push down on the cover with your thumb, the vial will sink to the bottom: release the pressure and it comes to the top. The explanation of the action of this miniature submarine is that pressure applied to the lid is transmitted to the water, forcing slightly more water into the vial. Its overall density is then just greater than that of water, and it sinks. Releasing the pressure allows the air in the top of the vial to push the extra water out again and the vial rises.

Air Resistance

So far we have been talking about air (gases) at rest. When air moves, a new force called air resistance begins operating. **Resistance** is the capacity of something to impede motion. You can feel the resistance of air by holding your hand out the car window. This air resistance is also exerted on the car itself; at usual driving speeds, more than half the power delivered by the engine may be used in working against air resistance. The gasoline "crises" and the increasing cost of fuel

Everyday Physics: 8

The colorful hot-air balloons that are currently popular are able to fly because heated air is lighter than cold air. Propane gas burners are used to heat air within the balloon. The air expands to fill out the balloon sac. As it becomes hotter and hotter, it becomes lighter and lighter until it can float on the cooler surrounding air and eventually lift the basket and its occupants above the earth's surface. Continuing bursts of flame are used to keep the air heated and the balloon aloft.

have made people more aware of this fact; when manufacturers advertise a car's low "coefficient of air drag" they are touting its relatively low air resistance.

The actual resistance force increases with the speed of a moving object and its cross-sectional area. In addition, shape is very important. What we call **streamlining** means designing an object to a shape that will offer minimum resistance to the flow of air past it at the speeds it is expected to travel. All sharp corners and projections must be eliminated. The overall shape should be that of a teardrop at ground speeds; contrary to what you might expect, the body should be broader toward the rear (Figure 8.4a). For a jet plane or rocket traveling faster than sound, a sharp-nosed shape gives the lowest air resistance (Figure 8.4b).

The air resistance of an object depends on its speed, its cross-sectional area, and its shape.

Figure 8.4

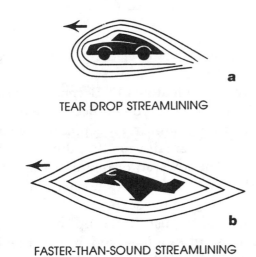

TEAR DROP STREAMLINING

a

FASTER-THAN-SOUND STREAMLINING

b

Figure 8.5 shows the comparative air resistance of three shapes: streamlined (teardrop), round, and flat, all having the same cross-section and all moving at the same speed. Behind the round and flat objects, the airstream breaks up into whirls that retard the movement of the object. The tapered "tail" of the teardrop fills in this region so that the flow of air moves smoothly to join the air to the rear of the object.

Figure 8.5

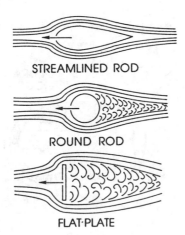

STREAMLINED ROD

ROUND ROD

FLAT·PLATE

When an object falls through the air, air resistance slows its fall. Area and shape affect air resistance, and that is why different objects fall at different rates.

Exploration 8.9

Drop a coin and a sheet of paper from shoulder height at the same instant. The coin quickly reaches the floor, while the paper flutters down slowly. To show that this result is not due to their difference in weight but only to the difference in air resistance, repeat the trial after first wadding the paper up into a small ball. This time both will be seen to hit at the same instant.

Bernoulli's Law

What keeps an airplane up? You may know in a general way that some form of air resistance keeps an airplane flying, but how? Whether we look at birds or airplanes, we can see that a large, slightly inclined surface—a wing—supports them in the air. If you have flown through light clouds, you may have been able to see how air streams backward, over, and around the wings. The tilted wing surface deflects some air downward so that the plane is literally "knocked" upward. Actually, it is the flow of air around the curved *upper* surface of the wing that accounts for most of the lift.

To see how this works, try this: Hold the edge of a strip of paper against your chin, just below your lower lip, in such a way that the paper curves up from your finger

Bernoulli's law: A moving stream of gas or liquid exerts less sidewise pressure than if it were at rest.

and then down. Now blow above the paper. It will rise to a horizontal position as if pulled upward into the airstream. You have demonstrated a principle known as **Bernoulli's law:** A moving stream of gas or liquid exerts less sidewise pressure than if it were at rest.

In a properly designed airplane wing, the airstream separates at the front of the wing and rejoins smoothly at the rear (Figure 8.6). The air that flows over the upper surface must travel farther, so its average speed is greater than the speed of the air below. As a result, a reduction in sidewise pressure occurs (Bernoulli's principle) at the top side, exerting a lifting force on the entire wing. The forces on the upper side of the wing may account for over 80 percent of the entire lift. The control surfaces on the wing (the flaps) operate on this same principle.

In a helicopter, the whirling of the rotors, rather than the helicopter's forward movement, provide the airflow over the wing (rotor) surfaces. As a result, a helicopter can hover over the ground or even fly backwards.

Other applications of Bernoulli's principle include everyday objects like atomizers. In an atomizer, air is pumped across the end of a small tube that dips into the liquid. The lowered pressure at the sides of the airstream allows normal air pressure, acting on the surface of the liquid in the bottle, to push liquid up the tube. As it comes out of a hole, moving air breaks it up into small drops and drives it forward.

A pitcher throwing a curve ball and a golfer hitting a "slice" are both demonstrating Bernoulli's principle (Figure 8.7). When a curve ball is thrown, the air on one side of the ball (A) is moving with the stream of air caused by the ball's motion; on the other side (B), the airstreams oppose each other. The greater relative speed at A makes the ball veer to that side.

Figure 8.7

Figure 8.6

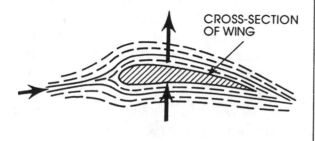

CROSS-SECTION OF WING

BALL VEERS THIS WAY

AIR DRAGGED AROUND BY SPIN

SPIN

Testing Your Knowledge

8.1 Explain why "water seeks its level"—that is, why the surface of a liquid at rest is flat and horizontal.
Hint: What would happen if the liquid were "heaped up" momentarily at one point?

8.2 Why are the hoops on a wooden water tank placed closer together near the bottom of the tank?

8.3 A dam or dike is made thicker toward the base. Explain. (See figure below.)

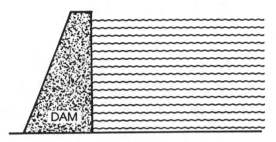

8.4 If there is a small hole in a dike at a point 10 feet below the water surface, does it take a greater force to keep the hole closed if the body of water is the Atlantic Ocean than it does if it were a small salt water pond? Why?

8.5 The water in an aquarium is 30 centimeters deep. What is the pressure at any point on the bottom?

8.6 The deck of a submarine is 100 feet below the surface of the sea (salt water). What is the pressure, and what is the total force, in tons, if the area of the deck is 1500 square feet?

8.7 A 100-lb sack of cement rests on a floor, making contact over an area of 80 in². The pressure, in lb/in², between the bag and the floor is about
a. 0.8.
b. 8,000.
c. 1.25.
d. 4.00.

8.8 The pressure at a point 5 ft below the surface of a pond
a. depends on the total depth of water in the pond.
b. is less than the pressure at a depth of 4 ft.
c. is greater than the pressure 5 ft below the surface of a gasoline storage tank
d. depends on the direction in which it is measured.

8.9 In a city water system, the water will flow
  a. only if the outlet is higher than the water in the standpipe.
  b. from the mains to the standpipe.
  c. faster from a first floor faucet than from one on the third floor.
  d. only when the standpipe has been completely emptied.

8.10 An open tank is shaped like a bucket. The diameter at the top is 10 ft; at the bottom, 8 ft; the tank is 6 ft deep. When brim full of gasoline, the pressure at the bottom will be, in lb/ft², about
  a. 7.3.
  b. 44.
  c. 264.
  d. 302.

8.11 Will aluminum, lead, and gold all float in mercury? Enumerate.

8.12 Explain the action of a cork life-preserver in terms of Archimedes' law.

8.13 When a ship sails out of a river into salt water, will the position of the water line on the side of the ship change? In what way?

8.14 A ferry boat has a cross-section area of 5000 ft² at the water line. How much lower will it ride, in fresh water, when a 20-ton trailer truck comes aboard?

8.15 If the overall density of an object is a certain fraction of the density of a liquid, then it will be able to float with this same fraction of the volume of the body under the surface. This being so, look up the densities of ice and of sea water (see Table 2.3) and decide what fraction of an iceberg is under water.

8.16 Knowing that normal atmospheric pressure can hold up a column of mercury 30 in high, use the relation $p = hD$ to prove that the pressure amounts to 14.7 lb/in². (The density of mercury as given in Table 2.3 must be changed to pounds per cubic *inch* by dividing by 1728.)

8.17 If the pressure inside a can of "vacuum-packed" coffee is 5 lb/in², what is the total force pressing down on the lid, whose diameter is 5 in?

8.18 What pressure is needed to compress 100 ft³ of air at normal pressure into a volume of 7.35 ft³?

8.19 A weather balloon filled with hydrogen has a volume of 4000 ft³ when on the ground. The bag itself weighs 50 lb. What weight of instruments can it carry and just get off the ground?

8.20 When you inhale, do you "suck" air into your lungs? Explain.

8.21 Why is it difficult to remove a rubber suction cup from a smooth surface?

8.22 Does the height at which the mercury stands in a barometer depend on the cross-section of the barometer tube? Give a reason for your answer.

8.23 A toy balloon, partly filled with air, is clamped shut at the neck and put into a closed jar. If the air is now pumped from the jar, what will happen?

8.24 Explain the action of a parachute in slowing down the motion of a body falling in air.

8.25 Explain why two motor boats, moving side by side, will tend to drift together.

PART TWO

SOUND AND LIGHT

CHAPTER 9

Sound

KEY TERMS FOR THIS CHAPTER

acoustics

oscillation

reflection of waves

sonar

frequency

wavelength

hertz (Hz)

ultrasonic

Doppler effect

intensity

amplitude

decibel

acoustic interference

resonance

forced vibration

The science of sound and how it is perceived is called **acoustics,** from the Greek word for hearing. Both sound and light, which we will discuss in Chapter 10, occur in the form of waves. To start, then, we should know something about waves and how they act.

The Nature of Waves

If you drop a stone in a quiet pond, you will see a set of waves spreading outward in ever-widening circles from the point where the stone entered the water. The size of each circular ripple grows at a constant rate (Figure 9.1). However, dropping a stone into water would only produce a few crests and hollows. To create a continuous wave motion, you would have to introduce a regular series of repetitive disturbances, for example, by inserting some vibrating object like a tuning fork into the water.

Figure 9.1

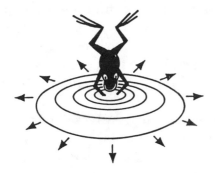

In any wave motion, no single particle of material carrying the waves ever moves very far from its normal place. If you were to watch a floating chip of wood in the water where you dropped the stone, you would see that it bobs up and down as each wave strikes it but does not move outward with each wave.

Different kinds of waves move particles in different ways. For example, particles moved by a compression wave, the kind that produces sound, **oscillate** (move back and forth about a center) along the line in which the waves are moving. For this reason they are called *longitudinal waves*. In other kinds of waves, the disturbed particles move perpendicular to the line of advance of the waves; these are called *transverse waves*. Still other waves are combinations of longitudinal and transverse motion; for example, the floating wood chip moves slightly forward and upward as the crest of a wave meets it, then back and downward as the next trough comes by.

When waves traveling on water strike a wide obstacle such as a floating board, you can see a new set of ripples starting back from the obstacle. The waves are said to be reflected from it. Sound waves and light waves can also be reflected, as you will see. You can observe the reflection of both water and light waves with a simple experiment.

Exploration 9.1

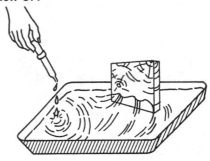

You can investigate the reflection of waves from flat and curved surfaces by using a large, flat pan with some water in it. Drop in some more water. The ripples you produce are easily seen if lighted by an unshaded lamp placed some distance above the pan. By adding a flat piece of wood or metal for the waves to hit, you can see how the waves will be reflected back.

As you watched the direct and reflected water waves, you may have noticed that the two sets of ripples can pass right through each other without having any mutual effect. This is true of any kind of wave, including sound waves. We will talk more about the effect of this phenomenon, which produces what are called *stationary wave patterns*, when we talk about sound.

Sound Waves

Suppose that, instead of tossing a stone into a pond, you were to explode a firecracker outdoors. You can demonstrate what would happen with an experiment of your own.

Exploration 9.2

Sound waves, which are also called compressional waves, are produced by a succession of compressions and expansions of the substance they occur in. To demonstrate, you could use a rubber balloon attached to a hand pump. First, push the pump handle down a short distance. The balloon will expand, quickly compressing the layer of air immediately around it. That layer of air will compress the next, and so on. The compression that started when you inflated the balloon will travel away from it in all directions.

If you then pull the pump handle up, the balloon will contract and the air around it will suddenly expand. This time, a region of low pressure will spread outward in all directions. If you move the pump handle up and down rhythmically (at regular intervals), a series of compressions and expansions will travel outward from the source. Such a regular series of disturbances is called continuous wave motion. If you could move the pump piston up and down fast enough, a nearby observer would hear the compressional waves as sound.

The sudden explosion of the firecracker compresses the air nearby. Air, being highly elastic, expands outward and compresses the layer of air just beyond. In this way the state of compression is handed on and spreads rapidly outward, just as ripples in a pond do. Here, however, we have a wave of compression, which is what a sound wave is. As it passes, the molecules of air crowd together, then draw apart. When the compression waves are sensed by nerves inside our ears, we hear.

Compressional waves can travel through any material, but they cannot travel in a vacuum. That is, they need some kind of matter as a carrier. Different substances carry sound at different rates. At room temperature (20°C), the speed of sound in feet per second is:

air	1,126 ft/sec
hydrogen	4,315
carbon dioxide	877
water	4,820
iron or steel	16,800
brass	11,500
granite	12,960

Notice that the speed of sound in air amounts to almost 770 mph, that in water sound travels almost four times as fast, and that in iron or steel it travels nearly 15 times as fast as in air. By knowing the speed of sound, you can estimate your distance from its source.

Exploration 9.3

You can estimate how far away a thunderstorm is by counting the number of seconds between the lightning flash and the thunder clap that accompanies it. You see the flash almost instantly because it is carried by light waves that

---

Sound can travel through any material but cannot travel in a vacuum (a space that contains no matter).

Sound waves are caused by alternating compression and expansion of air and are also called compression waves.

---

travel about 900,000 times as fast as the sound waves that bring you the noise of thunder. Since sound waves take about 5 seconds to go a mile, simply divide the number of seconds' delay by 5 in order to get the distance in miles.

The speed of sound depends on the temperature of the substance through which it is passing. For solids and liquids the change is usually small and can be ignored when making calculations, but for gases, including air, the change is large. Speed increases with temperature. At moderate temperatures, the increase is about 2 feet per second for each degree C. For example, the speed of sound in air at 20°C is 1126

feet per second; at 39°C the speed would be:

1126 + (the difference in °C × 2) = V (speed)

or

1126 = (19 × 2) = 1164 ft/sec

In still air that is at the same temperature throughout, sound travels uniformly in all directions, but of course this condition rarely exists. On a hot summer day the air next to the ground is hotter than the layers above. Since the speed of sound increases with temperature, sound will travel faster near the ground so that the waves will be bent away from the surface as in Figure 9.2. These waves will not reach an observer at P, so we say that sound does not travel far on hot days.

EVERYDAY PHYSICS: 9

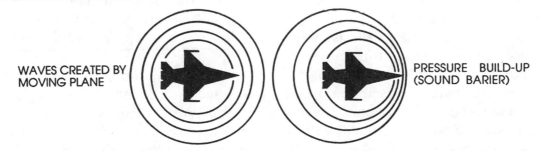

WAVES CREATED BY MOVING PLANE

PRESSURE BUILD-UP (SOUND BARIER)

As a plane travels through the air, it produces waves that travel ahead of it. As the plane reaches the speed of sound, it begins to catch up with those waves ahead of it. As the waves the plane produces as it increases speed get closer and closer to these waves it has already produced, the pressure in the air builds up to form a barrier—the sound barrier. When the plane exceeds the speed of sound and passes through this pressure barrier, the sudden increase in pressure results in a "sonic boom."

Figure 9.2

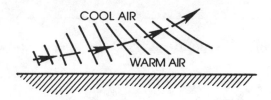

Figure 9.3

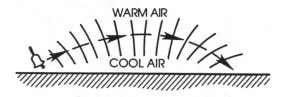

On a clear night, the ground cools more rapidly than the air above, so sound travels faster at some distance from the earth than it does nearer the surface. This has the effect of bending the waves toward the earth as in Figure 9.3, and the sound carries farther than usual.

Wind can cause similar effects. Wind speed usually increases with altitude so that sound waves will be bent toward earth and consequently can be heard farther away. The speed of sound waves going upwind (against the wind) will be slowed—the higher the altitude, the greater the slowing of the wave. These waves will be refracted upward and will be audible for a shorter distance.

Reflection of Sound Waves

You already observed how waves of water are reflected (turned back) by obstacles. Sound waves can be reflected in a similar way by mountains, walls, the ground (remember that sound waves travel in all directions), and even low-hanging clouds. The prolonged "rolling" of thunder is usually due to successive reflections from clouds and ground surfaces. You can use a rope to make a mechanical model of the reflection of sound (Exploration 9.4).

Exploration 9.4

To make a mechanical model of the reflection of sound, tie one end of a heavy cord or rope to a doorknob. Holding the other end in your hand, pull the rope fairly taut and give your hand a sudden downward jerk. A "hump" will travel down the rope and be reflected from the fixed end, returning to your hand in the form of a "hollow." You may note several back and forth reflections before the wave dies out.

A series of sound waves and reflections will act like the reflected waves of water you observed in Exploration 9.1. They can pass through each other without having any mutual effect, which means that you can distinguish separate sounds—for example, the voice of a singer and the sounds of his guitar—even though they mingle in

---

The speed of sound increases as temperature increases; it also increases with wind speed (air motion).

the air. Go back to the rope you used in Exploration 9.4 to demonstrate what happens.

Instead of giving the rope a single snap, shake it up and down in a rhythmic pattern. You will be setting up two continuous wave trains along the rope: the direct set going toward the door and the reflected set coming back. At any instant, the vibration of any given particle of the rope is determined by the resultant of the two wave motions as they pass. At the two ends of the rope, which are held fixed, the motion is always zero.

By trial, you can find a rate of shaking the rope at which all semblance of movement *along* the rope seems to disappear. At that point you will have produced a *stationary wave pattern*. The rope will vibrate back and forth across its rest position in the shape of a single arch.

If you double the rate (frequency) of shaking the rope, you will again get a steady pattern. This time the rope forms two equal loops. *Its middle point remains fixed even though no outside force is holding it fast.* Multiply that frequency of shakes by 3, 4, 5, or any whole number, and you

will produce that number of equal loops and fixed points. You will see the importance of this when we discuss wavelength and frequency.

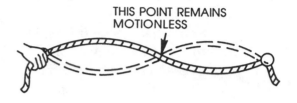

THIS POINT REMAINS MOTIONLESS

The human ear can distinguish two sounds as being separate only if they reach it at least a tenth of a second apart. Otherwise, the hearing mechanism of the ear blends them into the impression of a single sound. If a short, sharp sound is reflected back to you after more than about a tenth of a second, you will hear it as an echo—a repetition (reverberation) of the original sound.

The reflection of sound has a number of practical uses. Underwater sound waves are used in a device called a fathometer. (Water depth is commonly measured in fathoms, a unit of about 6 feet.) Sound is sent out in pulses from the device toward the sea bottom, and the reflected waves are detected by a receiver at some distance from the pulsar. The elapsed time between sending of the sound pulse and receiving the reflected sound is recorded, and the depth of the sea is calculated on the basis of the speed of the ship and the speed of sound waves in water.

The human ear distinguishes sounds as separate only if they are more than a tenth of a second apart.

An echo, or reverberation, is the repetition of a short, sharp sound reflected back after more than a tenth of a second.

Sonar (**so**und **na**vigation **r**anging) is used for underwater navigation and to locate underwater objects including schools of fish and sunken ships. Active sonar uses a "ping" or signal that bounces off anything it strikes (Figure 9.4). An instrument called a transducer converts the mechanical energy of the sound into electrical signals that are then converted into a pattern on a screen. Passive sonar does not emit a sound but detects sounds emitted by other objects. It is used in antisubmarine warfare, where it would be undesirable for the crew of an enemy submarine to detect the "ping" and thereby know they were being monitored.

Figure 9.4

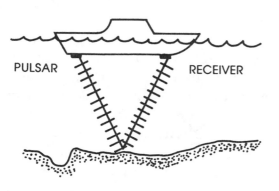

PULSAR   RECEIVER

Reflecting sound waves from hollow surfaces can increase their intensity in certain directions. That is one reason why auditoriums and amphitheaters are generally built in a concave shape rather than a square. In Statuary Hall in the Capitol in Washington, D.C., a person standing a few feet from the wall can hear the whis-

pering of another person who stands facing the wall at the opposite side, about 50 feet away. At intermediate points the sound is not heard.

Wavelength and Frequency

When a vibrating object sends out waves, the number of waves produced in one second is the same as the number of vibrations per second, that is, the **frequency,** of the source. The **wavelength** is defined as the distance between two successive places in the wave train that are in the same state of compression. In Figure 9.5, the strip across the top represents a sound wave, and the wavy line is a graph of the way the pressure in this wave changes. That is, the height of the curve at any point gives the pressure, above and below normal pressure, at that place in the wave train.

Figure 9.5

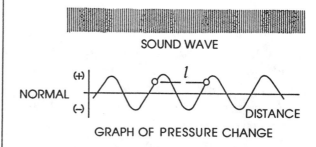

SOUND WAVE

GRAPH OF PRESSURE CHANGE

The distance l on this graph is one wavelength. It could have been measured from one crest to the next or from one trough to the next; the distance would be the same. The frequency n, the wavelength l, and the speed of the wave are related in such a way that:

n (frequency) × l (wavelength) = V (wave speed)

or

$$nl = V$$

Suppose the source vibrates for exactly one second. In this time, just n complete waves will be sent out (Figure 9.6). Each of the waves has the same length l, so that the first wave will be at a distance nl from the source and will have traveled a distance equal to V, the wave speed. The equation $nl = V$ holds for any kind of continuous wave, and it can be applied to find frequency, distance, or speed.

Figure 9.6

SOURCE   WAVE LENGTH
                                    WAVES

DISTANCE V

For example, in still air at ordinary temperature, the sound waves coming from a certain whistle are 27 inches long. What is the frequency of the sound?

In applying the wave equation, we need to use the same length unit throughout. Since the speed of sound is usually given in feet per second (ft/sec), we must change the wavelength from inches to feet: 27/12 = 9/4 feet. Then:

$$n = \frac{V}{l} \quad \frac{1126 \text{ (speed of sound in air)}}{9/4 \text{ (length of wave)}} \text{ (divided by)} =$$

500 vibration/sec

Exploration 9.5

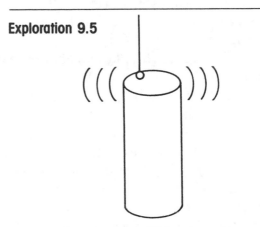

Hang a bead or small button from a piece of silk thread about 2 feet long, so that the bead just touches the lip of a large, thin drinking glass. Holding the glass firmly at its base, set it into vibration with a sharp snap of the finger. The glass will chime. Notice that the bead bounces aside violently as long as the sound lasts, showing that a sounding body is actually in a state of mechanical vibration.

Frequency and Pitch

We have different names for different kinds of sounds. The kind of sound we call a musical *tone* is heard only if the vibrations of the source, and therefore of the sound waves coming from it, have a definite frequency. Vibrations that are irregularly timed, and jumbles of unrelated

Frequency is the number of times per second that a sound source vibrates.

The frequency of a sound can be expressed in hertz (Hz), a unit referring to cycles per second, most commonly electromagnetic wave cycles.

---

vibrations, are called *noise*. The *pitch* of a sound—its degree of highness or lowness on the musical scale—is judged by the frequency of the sound waves. The faster the vibrations (meaning the higher the frequency), the higher the place of that tone on the musical scale. However, noise also has pitch: Draw your fingernail across a rough surface like a linen book cover. The sound produced is not very musical, but it does have a definite pitch. The faster you move your finger, the higher the pitch becomes.

Not all sound waves can be heard. A normal human ear can respond to frequencies from around 20 to just under 20,000 vibrations (sound cycles) per second. The frequency of sound is often stated in **hertz,** a unit referring to wave cycles per second. One **hertz** (Hz) equals one cycle per second; one kilohertz (kHz) equals 1000 cycles per second. For example, the range of sound a stereo amplifier can produce is stated in Hz; one capable of reproducing 20 to 20,000 Hz would be reproducing the entire audible range.

Frequencies above the hearing range of humans, especially those of several hundred thousand cycles per second, are called **ultrasonic** frequencies. The megahertz (MHz), which equals 1 million cycles per second, is generally used to express ultrasonic frequencies. Sounds at these frequencies can be produced by special methods such as electrically vibrated crystals or piezoelectric transducers in which crystals shrink when an electric field is applied to them.

Ultrasound can shake matter back and forth extremely fast. It can destroy bacteria in water, clean metals, remove smoke from air, and drill holes in hard or brittle materials. Ultrasound can be used to detect flaws in materials, trace the flow of coolants in nuclear reactors, determine the flow of blood in the body, and perform brain surgery.

The Doppler Effect

Have you ever noticed that, when an ambulance approaches and then passes you, the sound seems to drop off suddenly as the source sweeps past? When a listener and a source of sound are approaching each other, the waves strike more frequently than when both are relatively at rest. Therefore the tone seems higher than normal. When sound source and listener move apart, fewer waves hit the ear each second, giving the impression of a lower tone (Figure 9.7). This is called the **Doppler effect** after the scientist who first explained it.

Figure 9.7

WAVES SPREAD APART HERE - PITCH LOWERED

WAVES CROWDED TOGETHER HERE - PITCH RAISED

In astronomy, the Doppler effect on light waves is used to calculate the speed of stars relative to the earth. The Doppler effect on radar waves can be used to measure the speed of an airplane or satellite. Doppler radar is also used in airplanes to evaluate the danger of wind shear and in sonar systems to gauge the speed and direction of ships, submarines, and torpedoes.

Intensity (Loudness)

What makes sound loud or soft? One way to find out is by looking at it. Sound waves cannot be seen, but their patterns can be made visible with a device called an oscilloscope. The sound waves strike a microphone, which converts them into electrical disturbances that in turn are converted to images on a monitor similar to a TV screen. If you were to hold a gently vibrating tuning fork in front of the microphone, the soft sound would be translated into a pattern like this:

SOFT SOUND: SMALL WAVE AMPLITUDE

If you made the tuning fork vibrate more strongly to give off a louder tone (still at the same pitch), the patterns would look like this:

LOUD SOUND: LARGE WAVE AMPLITUDE

The wavelength and frequency would remain the same, but the crests of the waves would be higher and the hollows (troughs) deeper. In technical terms, the air particles would be vibrating at a greater **amplitude.** The greater the amplitude, the greater the **intensity** (loudness) of the sound.

Amplitude is not the only factor determining intensity. Distance from the source also affects intensity. If there are no disturbances such as reflections of the sound, intensity will diminish ("fall off") inversely as the square of the distance. In other words, at twice the distance, the intensity will be one-half squared, or one-fourth; at three times the distance, it will be one-third squared, or one-ninth, and so on.

The intensity of sound as received at any point is measured in a unit called the **decibel** (dB). In Table 9.1, which lists the intensity of common sounds in decibels at

---

The Doppler effect explains why sounds increase in pitch as a sound source nears but fall off sharply as it passes.

The intensity (loudness) of a sound depends on amplitude of the sound wave.

---

the distance from which they are generally heard, the differences in number look small. However, the range from 0 dB for the least sound perceptible by the average human ear to 130 dB for the amount of sound that produces pain represents a range in *energy* to a factor (self-multiple) of 100 billion.

Prolonged exposure to loud noises can permanently damage the ear, leading to hearing loss or deafness. Awareness of this has led to such developments as the sound baffles now used in highway construction. Devices such as stereo headphones, which deliver sound directly to the ear, have increased the risk of hearing damage.

Since intensity falls off so rapidly with distance, hearing unamplified speech or music outdoors is not very satisfactory.

Table 9.1   Sound Intensity Levels

Source	dB
Faintest audible sound	0
Rustling of leaves	8
Whisper	10–20
Average home	20–30
Automobile	40–50
Ordinary conversation	50–60
Heavy street traffic	70–80
Riveting gun	90–100
Thunder	110

The sound may seem flat and "dead." In an enclosed space, the reflection of sound from walls and other surfaces makes the intensity more nearly uniform throughout the space. These reflections also make the sound impression last longer by increasing the *reverberation time* of the sound. This gives the tones more "life." (That is why your singing sounds better in the shower.)

Not all spaces are equally good at producing pleasant sound. Hard surfaces such as floors, walls, and unupholstered furniture may reflect the sound waves back and forth too often before being absorbed. Excessive reverberation time causes tones to mingle unpleasantly with those that were previously produced. The shape of the room may cause "dead spots" and "hot spots"—places where the sound disappears or is too loud. Similar problems may arise in your home if your stereo speakers are not placed properly. When sound is pleasantly audible throughout a space, we say the space has "good acoustics."

Acoustic interference occurs when two or more sound waves are produced at the same time. Their amplitudes at any point in space may be added together to determine their combined effect (Figure 9.8). If the amplitude of the resultant wave is larger than either of the original waves, the waves are said to interfere constructively. If the resultant amplitude is

---

The intensity of sound is measured in decibels (dB).

---

smaller than one of the individual waves, the waves interfere destructively.

Figure 9.8

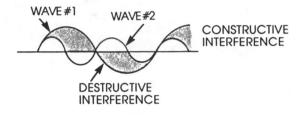

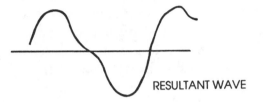

RESULTANT WAVE

Music: Controlled Sound

To produce the patterns of sound we call music, it is necessary to control, or limit, the frequency and intensity of the sound waves. Different types of musical instruments do this in different ways. Wind instruments such as the organ, flute, and trumpet do this by means of a property of sound known as **resonance:** the process by which sound vibrations build up.

Resonance reinforces sound in much the same way that a mechanical movement can be reinforced. Suppose you are pushing a swing. To get it to swing high, you must push it at the exact tempo of its natural frequency of motion. A slight force applied each time the swing reaches its highest point will soon build up a wide movement.

An example from acoustics would be the rattling of a window when a low-flying airplane passes overhead. If the natural frequency of vibration of the windowpane happens to be the same as one of the frequencies that make up the noise of the plane's engines, the window will reinforce (resonate) the sound. In the same way, a certain note on the piano may make the chandelier tinkle, and a singer's high note may cause a glass to vibrate to the point of breaking.

Exploration 9.6

Pinch together the prongs of a dinner fork so that they are set into vibration. The sound is very faint, but if you now press the end of the handle firmly down against the top of a hard table the sound at once becomes remarkably loud.

In a closed organ pipe (Figure 9.9a), a jet of air blowing just inside the sharp lip of the pipe builds up extra pressure there. This region of compression travels down the pipe at the speed of sound and is reflected back. When it reaches the open end once more, the compression pushes the airjet out of the tube. The pressure is relieved, the airjet comes back into the tube again, and everything repeats. The length of the pipe regulates the frequency of vibration of the air jet and thereby the frequency (pitch) of the note.

Exploration 9.7

The simplest of all wind instruments is a tube or tall bottle that you blow by directing a stream of air across the open end, just inside the far edge. Try bottles of different length and observe that the longer ones give deeper notes.

Figure 9.9

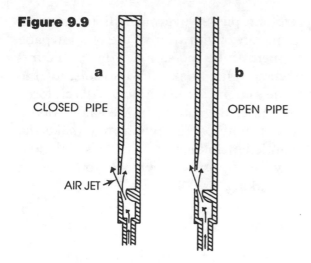

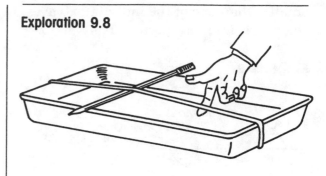

An open organ pipe as well as most brass and woodwind instruments are open at the far end. The way in which sound is reflected from the end is changed by this; it is still the length of the pipe that controls the pitch of the tone (Figure 9.9b). Other wind instruments use flexible reeds or the lips of the player instead of an air jet, and some, like the cornet and tuba, shunt air through different lengths of tubing.

A vibrating body may also transmit its movement to another body with a large surface to magnify the sound by setting more air in motion. The sounding board of a piano, the body of a violin, and the diaphragm of a loudspeaker are all examples of **forced vibrations**: they amplify *all* frequencies transmitted to them.

Stringed instruments produce sound by the principle you observed with the rope experiment. Bowing, plucking, or striking the string at different points in its length creates different frequencies of vibration. The thickness of the string and the point at which it is stopped or struck are ways of controlling the frequency of the sound produced.

Check some of the facts about vibrating strings by making a simple one-stringed guitar. Stretch a rubber band around a long pan, such as a baking tin. Use a stick or pencil as a "bridge." With the bridge absent, pluck the string and note the pitch of its tone. Then insert the bridge under the center of the string, pluck either half, and notice that the tone produced is the octave of the first one. Placing the bridge at the quarter point will give the next octave. Try also to produce, one after another, the familiar *do-mi-sol-do* of the major chord by using the "open" string, then placing the bridge at distances $\frac{1}{5}$, $\frac{1}{3}$, and $\frac{1}{2}$ from the left end, each time plucking the right-hand portion of the string. Another fact you can check is that tightening a string raises its pitch.

Percussion instruments such as drums and bells create tones by the striking of bars, plates, stretched skins, or some other material that give their motion to the surrounding air. The thickness of the material, its density, and so on determine its rate of vibration and therefore the pitch of the sound.

Different types of instruments also produce different qualities of sound. These have to do with harmonics, also called overtones, which are vibrations at different strengths produced by different ways of producing the tone. Electrical devices called harmonic analyzers can analyze the complex waveforms produced by different sources (Figure 9.10), and modern electronics can be made to reproduce (synthesize) these natural waveforms.

Figure 9.10

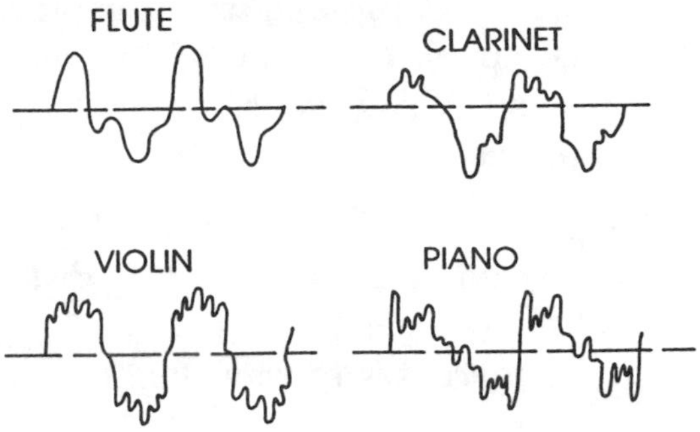

Testing Your Knowledge

9.1 Explain why soldiers marching near the end of a long column are observed to be out of step with the music of a band marching at the head of the column.

9.2 Give a reason for believing that sounds of different pitch, such as those coming from various instruments of an orchestra, all travel at about the same speed in air.

9.3 A sounding device on a ship shows an echo coming back after 3.5 seconds. How deep is the water if the speed of sound waves in sea water at the existing temperature is 4700 ft/sec?

9.4 If the earth's atmosphere extended uniformly as far as the moon, how many days would it take sound to travel that distance? On the average, the moon is about 239,000 mi away.
  a. 310
  b. 13
  c. 665
  d. 25

9.5 A wave in which the particles of the material move up and down as the wave goes from left to right is classed as
  a. longitudinal.
  b. transverse.
  c. compressional.
  d. sound.

9.6 On a certain day, sound is found to travel 1 mile in 4.80 seconds. What is the temperature of the air through which the sound is passing?
  a. 7°C
  b. 27°C
  c. 26°C
  d. −6°C

9.7 How long are the sound waves produced in the air when middle "C" is struck on a piano, the frequency of vibration being 256 vibrations per second?
  a. 0.23 ft
  b. 2.3 ft
  c. 288 ft
  d. 4.3 ft

9.8 In the last problem, some of the sound is allowed to pass into a tank of water. Compared with the wavelength in air, the length of the compressional waves in water will be
a. slightly greater.
c. about 4 times as large.
b. the same.
d. considerably less.

9.9 In older fire sirens, a jet of air is directed against a series of evenly spaced holes in a rotating disk. As the disk speeds up, the tone
a. increases in frequency.
c. drops in pitch.
b. increases in wave length.
d. maintains constant pitch.

9.10 In order to emit sound, a body must
a. absorb sound waves.
c. reflect sound waves.
b. vibrate.
d. move toward the hearer.

9.11 As a man moves directly away from a steady source of sound at constant speed, the sound he hears will
a. increase in frequency and intensity.
c. decrease in frequency and intensity.
b. stay constant in pitch but decrease in loudness.
d. remain constant in both pitch and loudness.

9.12 To a man who moves from a position 20 ft from a bell ringing steadily in the open air to a position 60 ft from the bell, the intensity of the sound will appear to
a. decrease by a factor of 3.
c. increase by a factor of 10.
b. decrease by a factor of 9.
d. decrease by a factor of 1,200.

9.13 A concert hall has too long a reverberation time. Of the following, the best way to correct the condition would be to
a. limit the size of the audience.
c. lay heavy carpeting on the floor.
b. install a curved ceiling.
d. hang large mirrors on the walls.

9.14 Explain the difference between echo and reverberation.

9.15 Middle C on the piano has a frequency of 256 vibrations per second. What is the frequency of the note one octave below this? One octave above?

9.16 Why do the bass strings of the piano have heavy, flexible wire wound around them along their entire length?

9.17 When water is poured into a tall jar, the pitch of the tone produced rises as the jar fills up. Explain.

9.18 Can you classify the human voice and the piano as wind, string, or percussion instruments?

CHAPTER 10

Light

KEY TERMS FOR THIS CHAPTER

umbra
penumbra
luminous intensity
standard candle
foot-candle (ft-c)

quantum
law of reflection
diffuse reflection
fiber optics
refraction

index of refraction
law of refraction
lens
telescope
microscope

Look around you. What do you see? A window, perhaps, with a scene beyond it. The setting sun or a street lamp. Your room, lamp, chair, this book. Actually, you are seeing rays of light sent out by luminous objects such as the sun or the light bulb and reflected by other objects such as furniture, people, and books. This chapter discusses the physical character of light and light sources; Chapter 11 explains how different kinds of optical instruments give us information about light.

Light Rays

When people began studying light, one of the first things they noticed was that it travels in a straight line. Have you seen shafts of sunlight coming through a rift in the clouds? Have you sighted along the side of a used car to see whether there are any dents? When driving at night, have you seen how your car headlights beam straight ahead, so that it's hard to see around a curve? All these things are evi-

---

Light travels in straight lines that we call light rays.

---

dence that light travels in straight lines called light rays.

If you stand by a wall with the light coming from behind you, you will cast a shadow on the wall. Imagine straight lines drawn from the light source and passing through the extreme outward edge of your shadow, which will be very sharply defined. These lines depict light rays (Figure 10.1). They may be drawn outward from the source in any number and in all directions to serve as a convenient guideline for showing the direction that light is traveling at any point. For you to see an object, rays coming from it must enter your eye.

Figure 10.1

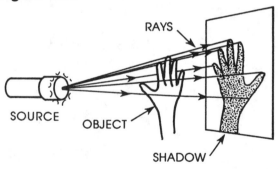

If a source of light is very large, every point of it must be thought of as a source of spreading rays. The result is that its shadow will have a central black part called the **umbra** and a lighter outer part called the **penumbra.** Inside the central shadow area, no light is received from any part of the light source. At the outer edges, some light is received. This accounts for what happens in an eclipse of the sun, when the moon comes squarely between it and the earth (Figure 10.2).

Figure 10.2

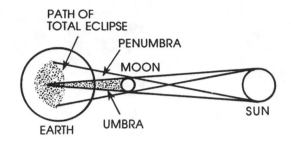

Notice how an observer who happens to be in the central region experiences a total eclipse (complete blocking out of the sun). Someone in the outer region sees a partial eclipse.

Exploration 10.1

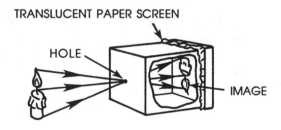

Make a pinhole camera, which utilizes the straight-line travel of light. Cut out one end of a cardboard box and cover this end with a piece of translucent paper. Make a clean hole in the center of the opposite end of the box with a darning needle. In a darkened room, a candle flame or light bulb placed a

For you to see an object, the light rays it sends or reflects must enter your eye; since light rays travel only in straight lines, you cannot "see around a corner."

few feet in front of the hole will be represented as a bright, inverted image on the wax paper screen. Notice that the closer you bring the object to the camera, the larger its image will be. Surprisingly good pictures can be obtained with a pinhole camera by putting a photographic film in place of the screen. But because of the small amount of light that can come in through the pinhole, the exposures are inconveniently long.

Artificial Sources of Light

Here on earth, the sun is our main source of natural illumination (the moon reflects the light of the sun). Human activities were restricted largely to the hours when the sun was above the horizon. No wonder people began, very early, to devise forms of artificial illumination. From their crude torches and oil lamps, we have come to a time when artificial illumination can flood huge stadiums with light.

The familiar electric light bulb is a *filament lamp*. Light is produced by a current of electricity passing through a very fine tungsten wire, or filament, placed inside a glass bulb that either contains a vacuum or has been filled with some inert gas such as nitrogen to prevent the filament from burning through. The current raises the temperature of the filament to around 2500°C, where it becomes "white hot." A white coating inside the glass diffuses the light and "softens" it. (Diffusion is explained later in this chapter.)

A *carbon arc* is a very intense source of light used in movie projectors and searchlights. Two carbon rods are connected to an electric battery or generator. The tips of the rods are brought into contact and then drawn apart a short distance, producing a flame of burning carbon that carries the electric current across the gap. Most of the light comes from the glowing tips of the carbons, which reach a temperature of 3000°C to 3500°C.

The colorful lighted signs we call neon signs are a common example of how a current of electricity flowing through a gas or vapor at very low pressure can create light. The tubes contain different kinds of gases, sometimes in addition to easily vaporized substances such as sodium or mercury. In this kind of lamp, the atoms of the gas directly convert electrical energy into light; there is no burning (combustion).

In fluorescent lamps, electricity passing through a mixture of argon gas and mercury vapor produces ultraviolet light.

The strength of a light source, its luminous intensity, is measured by the standard candle (c); today, however, luminous intensity is generally measured by comparison with a standardized filament lamp.

Illumination is a derived unit describing the amount of radiant energy that falls on a specified unit area; it can be computed by a formula in which *E*, the radiant energy, equals the inverse square of the distance from the light source.

Ultraviolet light is not visible, but it causes a chemical coating inside the tube to glow intensely.

More recently, a type of illumination called discharge lighting has been developed. These lamps also produce light by a combined action of a gas and a filament. The sodium vapor lamps used on expressways and the halogen headlights now seen in some cars are examples of this kind of lighting. Other light sources include light-emitting diodes (LEDs) and electro-luminescent panels like those used in instrument displays.

How Illumination Is Measured

The strength of a lamp or other light source is specified by a quantity called **luminous intensity,** which is measured by the **standard candle.** As the name suggests, this unit goes back to the ordinary wax candle as a source of light. Today, though, lamps are rated by comparing them to standardized filament lamps kept in testing laboratories such as the bureau of standards. A filament lamp of moderate size has an intensity of about one candle for each watt rating—a 60-watt lamp is almost 60 candles.

Illumination is the radiant (light) energy falling on any given unit area. Like sound, light spreads out in waves from its source. The farther an object is from the source of light, the larger the area over which a given amount of light will spread. This area increases as the square of the distance: At twice any given distance, a given amount of radiant energy will be spread over four times the distance. At three times the distance, it will be spread over nine times the area, and so on (Figure 10.3).

Figure 10.3

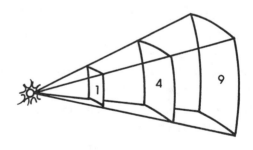

This means that the illumination, or radiant energy falling on each unit area, will vary inversely as the square of the dis-

---

The human eye is so sensitive to light that it can perceive the equivalent of the illumination produced by a single candle 20 miles away.

Table 10.1 Illumination Required for Different Purposes (in Foot-candles)

Comparable with outdoor light. For hospital operating rooms, TV cameras	1000
Color matching in manufacturing	500
Extra-fine inspection work	200
Prolonged seeing tasks involving fine detail	100
Lengthy seeing tasks such as drafting or sewing	50
Rough bench- and machine-work; reading or writing for short periods	20
Intermittent visual work; rough manufacturing, stockrooms	10
Auditoriums, corridors, general illumination in the home	5

tance from its source. The illumination of any surface that is held perpendicular to the incoming rays will also depend directly on the strength of the source. The complete relation is given by:

$$E \text{ (radiant energy)} = \frac{C}{d^2} \begin{array}{l}\text{(intensity of source)}\\\text{(divided by)}\\\text{(distance squared)}\end{array}$$

Illumination is a derived unit. If intensity (C) is measured in candles and distance (d) in feet, E (radiant energy, or illumination) is expressed in a unit called a **foot-candle.** The corresponding metric unit is called a meter-candle. A surface receiving 1 foot-candle of energy is getting energy at the rate of about 0.00001 watt per square centimeter (cm²). The human eye is so sensitive that its receptor nerves can be stimulated by as little as a *ten-billionth of a foot-candle,* which is equivalent to the illumination produced by a single candle nearly 20 miles away.

Now, try putting your new knowledge to use by computing where you need to place a lamp with a 60-c bulb to provide 15 foot-candles (ft-c) of illumination to the book you are reading. Using E (15 ft-c) = C (60 ft-c)/d^2, you get d^2 = 60/15 = 4 and d = 2 (the square root of 4).

The intensity of a lamp is usually measured by comparing the illumination it produces with that from a standard lamp by means of devices such as photometers. Engineers and architects have devices that can directly measure the total illumination of any surface. Some standard recommended illumination values for specific uses are listed in Table 10.1.

What Is Light?

What is light? Is it, as the Greek philosopher Plato thought, a substance that the eye gives off? Is it particles shot off by light sources? Is it something that happens in between the source and the object struck by the rays? As early as the seventeenth century, several scientists were independently suggesting that light is a wave motion. Isaac Newton considered this possibility, but concluded that light is a stream of particles darting out from luminous bodies. This corpuscular (particle) theory of Newton was accepted, despite evidence to the contrary, for nearly a century after his time.

If light is a wave, as physicists later concluded, what is it a wave *in?* Knowing that

---

Current theories of light combine Einstein's quantum theory, which conceives of light as existing of particles called photons or quanta, and Maxwell's theory of light as consisting of electromagnetic waves.

---

other kinds of waves travel—in water, in air, in strings—physicists conceived of a carrier they called "ether," which was supposed to fill all of space, carrying the light from the sun and the stars. The existence of ether could not be proven. Instead, the British physicist James Clerk Maxwell suggested in the mid-nineteenth century that a light wave consists of rapidly changing electric and magnetic forces that originate in the movement of electricity in the atoms or molecules from which the light comes. (These electromagnetic waves are discussed more thoroughly in Part Three, Chapters 12 through 16.) Early in the twentieth century, Albert Einstein worked with the concept that light is made up of little bundles of energy called photons or **quanta** (singular, quantum). The energy in a quantum is determined by its frequency. Today, a combination of wave theory and Einstein's quantum theory is accepted as being correct. The information about light presented in this chapter can be explained in terms of either wave behavior or particle behavior.

When you look at a steady light, you get no sense of its consisting of waves, any more than hearing a steady tone gives a sense of the wavelike nature of sound. We cannot see individual light waves, but must accept wave theory on the basis of indirect evidence. The facts about light discussed in this chapter could be interpreted by either the particle theory or the wave theory.

The Speed of Light

Like any wave, light takes some time to travel from one place to another. The Danish astronomer Olaus Roemer, working in the seventeenth century, calculated the speed of light by timing the appearance of one of the moons of Jupiter at different times in the earth's orbit. His calculation was the equivalent of what we now know to be the speed of light, or roughly 186 million miles per second.

In modern laboratories, speed of light is measured by how fast a beam of light covers a known distance. When you consider that light travels nearly 1000 feet in one-millionth of a second, you can see how difficult making such measurements can be. The best present calculation is that c, the standard symbol for the speed of light in a vacuum, is:

183,310 miles per second
299,776 kilometers per second

Light, unlike sound, can travel in a vacuum because it is not a wave resulting from the vibration of any medium but rather a series of moving magnetic fields (see Part Three).

In any transparent material, such as air, the speed of light is *less* the c. For example, light travels:

In air: $c - 0.03$ percent
In water: $c - 25$ percent
In glass: $c - 35$ percent.

Reflection

So far, we have been talking about light traveling in a uniform material. What happens if light comes to a non-uniform region or strikes a surface? Like any other kind of wave, it changes its direction or is reflected back.

In Exploration 9.1 in Chapter 9, you saw how circular ripples were reflected back from a flat obstruction in the form of ripples traveling in the opposite direction. The ways in which light is reflected from any object are governed by certain laws. We can study those laws by representing light rays as lines on a drawing (Figure 10.4).

Figure 10.4

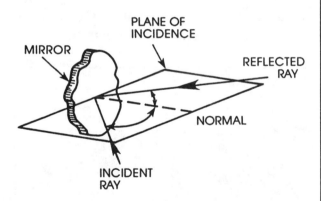

When light is reflected from a plane (flat) mirror, the incoming ray of light, called the *incident ray*, and the reflected ray of light, called the *reflected ray*, are measured with respect to the *normal*, a line perpendicular to (at right angles to) the plane surface. When any light ray strikes a plane, the angle of incidence always equals the angle of reflection, and the normal and the incident and reflected rays always lie in the same plane (law of reflection). However, the two rays lie on *opposite sides of* the normal; that is, the light goes "straight on."

By applying the **law of reflection** to particular sets of rays, you can work out the reflection of light from any kinds of surface.

At a considerable distance from a small source of light, a limited portion of the spherical waves will be practically plane. The rays, which are always perpendicular to the light waves, will in this case be nearly parallel (Figure 10.5).

Figure 10.5

Sunlight is an example. If a parallel beam of light hits a plane mirror, the law of reflection tells us the rays will also be parallel after reflection (Figure 10.6). This is called *regular reflection*.

The law of reflection states that the angle of incidence (*i*) always equals the angle of reflection (*r*), and that the angle of incidence, the angle of reflection, and the normal all lie in one plane.

Figure 10.6

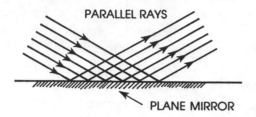

PARALLEL RAYS

PLANE MIRROR

By contrast, when a beam of light strikes a rough or irregular surface, the angles of incidence and reflection will be equal at *each point* on the surface, but the various portions of the surface have different directions, and so do the reflected rays. As a result, a rough surface will be visible from almost any position, while, to receive light from a mirror, your eye must be in the particular direction in which the incident beam is reflected. Reflection from a rough or irregular surface is called **diffuse reflection** (Figure 10.7).

Figure 10.7

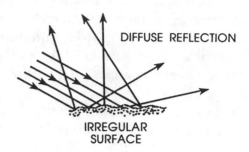

DIFFUSE REFLECTION

IRREGULAR SURFACE

We recognize material objects from the effect of their surface texture on the light they diffusely reflect. A perfectly smooth, clean mirror would be invisible; you would see the source of the light rather than the mirror.

A highly polished silver surface reflects about 95 percent of the light that falls on it perpendicularly. An ordinary mirror, consisting of a sheet of glass silvered on the back, reflects about 90 percent. Even

---

EVERYDAY PHYSICS: 10

Light can be "piped" along a rod of transparent plastic material by total reflection. In surgery and dentistry, this method provides concentrated light without objectionable heat and allows exploring the insides of blood vessels and other internal organs. This is the principle of **fiber optics.** Glass or plastic is drawn out in very fine fibers. Light can be transmitted through the fibers over very long distances. The light is reflected off the inside faces of the fiber, from one end to the other. Lasers as small as a grain of salt are being developed to transmit signals through optical fibers, which already carry millions of telephone messages over long distances.

without being silver-backed, a transparent surface such as a sheet of glass reflects some light. Have you ever sat in a lighted room at night and noticed how the room's interior is reflected back by the windows? An unsilvered sheet of glass reflects only about 8 percent of the light falling perpendicularly on its surface, 4 percent from the rear surface, 4 percent from the face. At large angles of incidence called *grazing incidence*, however, almost all of the incoming light is reflected at the front surface. That explains why the sun's reflection in a lake is not extremely bright when the sun is overhead but too dazzling to look at when the sun is low in the sky.

Mirrors

When we say we look *into* a mirror, we are subconsciously recognizing the fact that light originating from any point in front of a mirror will *appear* to come from a point at equal distance behind the mirror. Actually, nothing is going on behind the reflecting surface, but the point from which the light seems to be originating is called the image of the source (Figure 10.8).

Figure 10.8

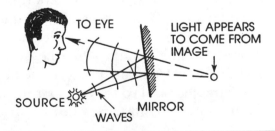

Because the light rays only seem to originate from this point, the image is said to be *virtual*. You can find the image of a real object of any size by taking one point after another and locating its image. You will find that the complete image is the same size as the object, and the positions of the object and image are symmetrical with respect to the mirror (Figure 10.9).

Figure 10.9

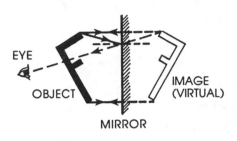

When you look at yourself in a plane mirror, right and left are reversed. If you wink your right eye, the left eye of the image winks back. (Photographers have found that people tend not to recognize their photographic images if the right and left sides of their faces are noticeably different—as most people's are. For this reason, some professional photographers "flop" the negative so that the finished photograph looks like the person's mirror image—the only one he or she ever sees.) When any object is reflected in a mirror, object and image are identical in all detail but arranged in reverse order.

How large must a mirror be for you to see all of a given object in it? Even a tiny

---

Any object reflected in a plane mirror will be identical in all details but arranged in reverse order; a "mirror image" runs "backward," as you can see by holding this book in front of a mirror.

mirror will reflect the entire image, but *you* may have to move around in order to see all parts of it. (See Exploration 10.2.)

Exploration 10.2

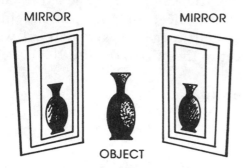

Stand two mirrors up a few inches apart so that they face each other and are parallel. Stand a small object like a lipstick case between them. When you sight over the top edge of one of the mirrors you will see an endless row of images in the other. Any image formed in one mirror acts as the object for another image formed in the other mirror, and this goes on indefinitely. There should be a similar row of images in each of the two mirrors, as you can check by looking in from the opposite side.

Refraction

The action of mirrors depends on the fact that light hitting a surface is turned back into the space from which it comes.

This always happens to some extent, but if the surface is made of a transparent material, some of the light passes through into the second substance. If the light comes in at an angle to the normal, it will change its direction sharply in going through the boundary. This change in the direction of rays is called **refraction.**

If a ray of light passes from a less dense substance such as air into a denser substance such as glass, it will be bent *toward* the normal at the surface of separation. If the ray is passing from a denser to a rarer (thinner) material, it will bend *away from* the normal. The amount of deflection (bending) will depend on the angle of the incoming ray (Figure 10.10).

Figure 10.10

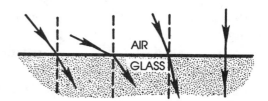

Have you looked down at your legs while standing in shallow water? Or at a spoon standing in a half-filled glass? Both seem to take a sharp bend where they enter the water. This is because any underwater object is seen by light reflected from it and coming up through the surface. Rays coming from any point in the object are refracted, and so they appear to come from another point closer to the surface.

Light rays passing through a transparent substance are bent, or refracted; the thicker the substance the farther apart its actual and apparent locations will be.

The angle where total internal reflection occurs is called the critical angle; at this angle, a light beam emerging from a transparent substance such as water will lie so close to the surface that it will no longer emerge but will be reflected back into the water.

The immersed part of the object seems to be swung upward.

When light passes completely through a piece of glass that has parallel sides, the rays will be bent equally at each surface, but they will be bent in opposite directions—toward the normal in one case and away from it in the other. The beam that comes through will be parallel to its original course but displaced to one side. You can see this for yourself: Hold a piece of glass at an angle to your line of sight and look at a pencil through it. The part of the pencil behind the glass will appear moved to one side.

Exploration 10.3

Light passing through air that is not all at the same temperature is refracted. This accounts for the twinkling of the stars and the wavering appearance of objects seen over a heated surface, such as the top of a car that has been standing in the sun. It also explains the occurrence of a mirage. Riding in a car on a day when the sun is shining but the air is cool, one often notices the presence of what seem to be pools of water on the road some distance ahead. The "water" always disappears before you get to it; what you see is merely light from the sky that is refracted to your eye by warm air near the ground. The effect is exactly what was represented for sound waves by Figure 9.1.

The observed direction of a refracted ray can be explained by the wave theory. Figure 10.11 shows a parallel beam of light coming through the air and striking a flat glass surface. AB is a plane wave front that is just about to enter the glass at point A. CD is a wave front that has just gotten completely in. During the time the light waves travel a distance da in the air, they evidently travel a distance dg in the glass. Taking V_g as the speed of light in glass and V_g as the speed of light in air, we can say that

$$\frac{V_a \text{ (speed of light in air)}}{V_g \text{ (speed of light in glass)}} = \frac{d_a \text{ (distance in air)}}{d_g \text{ (distance in glass)}}$$

For all practical purposes we can take V_a to be the same as c, the speed of light in a vacuum.

Figure 10.11

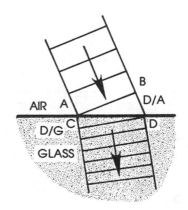

The ratio of light's speed in a vacuum to its speed in a given material is called the **index of refraction.** This quantity, represented by the symbol n, determines how much the rays are bent. Thus, we can view the **law of refraction** as giving us the change in direction of a ray in terms of the value of n and the construction of Figure 10.10. Remember that the law also states that the incident ray, the normal to the surface, and the refracted ray all lie in one plane. Table 10.2 gives the index of refraction of some common substances.

An interesting case of refraction occurs when light comes up toward the surface of water from below. Figure 10.12 shows an underwater spotlight that can be inclined at various angles. If the lamp is inclined more and more to the surface of the water, the emerging beam will lie closer and closer to the surface until, when the beam from the lamp makes an angle of about 49 degrees with the normal, the outgoing beam will go right along the surface. If the angle is made any greater than 49 degrees, there will no longer be an emerg-

Table 10.2 Index of Refraction of Various Substances

Substance	Index n (= c/V)
Air	1.0003
Ice	1.31
Water	1.33
Gasoline	1.38
Olive oil	1.47
Glass, ordinary	1.5
Glass, dense optical	1.6–1.9
Diamond	2.42

ing beam—all the light will be reflected back into the water. This is called **total internal reflection.** The angle this happens at is called the **critical angle.**

Lenses

Refraction finds its most useful application in **lenses,** which are an essential part of many optical devices such as microscopes, projectors, eyeglasses, telescopes, cameras, and rangefinders. The purpose of a lens is to change the curvature of light waves—that is, to bend the rays—usually in order to form an image. Lenses are ordinarily made of optical glass or plastic. In practice, the two surfaces are ground to spherical form. The commonest lens shapes are shown in Figure 10.13. Those that are thicker at the center than at the edges are called converging lenses, those thicker at the edges are diverging lenses. The reason for these designations will be explained.

Figure 10.12

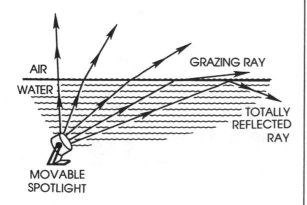

Figure 10.13

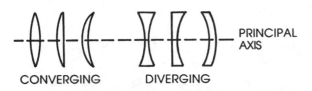

CONVERGING   DIVERGING

Figure 10.14

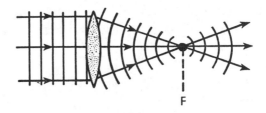

F

The ability of a lens to bring light to a focus is measured by what is called focal length. Consider a beam of parallel light (plane waves) coming in along the axis of a converging lens, as shown in Figure 10.14. Because light moves more slowly in glass than in air, the part of a wave that passes through the thick middle portion of the lens will be held back more than the outer parts. After getting clear through the lens, the waves (which were originally plane) will now be parts of spheres that close down to a point *F*, the principal focus of the lens. If there is nothing to stop them, the waves again expand as they go on. The distance of the principal focus from the lens is called the focal length of the lens. Its magnitude depends on the curvature of the two surfaces and on the index of refraction of the material of which the lens is made. The whole situation is reversible: If a point source of light is placed at the principal focus of a converging lens, the waves will be "straightened out" and made plane by passing through it.

How a Lens Forms an Image

Now that we know what a converging lens does to light coming from a great distance (plane waves), let us see what happens when a source of light is placed nearer to the lens, but at some point *P* beyond the principal focus (Figure 10.15a). The lens is still able to reverse the curvature of the waves, and they close down to a point *Q* that lies beyond the focal distance on the right side of the lens. However, if the object is placed anywhere *within* the principal focus, the lens can no longer change the direction of curvature of the waves, and there is only a virtual focus at some point on the *near* side of the lens. This is the case when a lens is used as a *simple magnifier*.

If the point source at *P* in Figure 10.15a is replaced by an extended light-giving object (Figure 10.15b), a complete image of this object will be formed at *Q*. Rays from each point of the object are brought to a corresponding focal point of the image.

The focal length of a converging lens, which is the distance of the principal focus from the lens, expresses its ability to bring light to a focus.

The image is real; rays actually cross to form it, and it can be caught on a screen. It is also inverted, and if the object extends in a direction perpendicular to the page, it will be reversed from side to side as well.

Figure 10.15

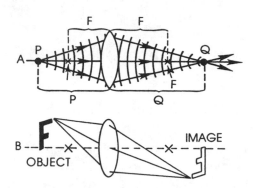

How can we find the location and size of a real image? The relation giving the distance is found to be

$$\frac{1}{p} + \frac{1}{q} = \frac{1}{f}$$

where p is the object distance, q the image distance and f the focal length of the lens. The size of the image is given by

$$\frac{h_1}{h_0} = \frac{q}{p}$$

where h_1 is the height of the image and h_0 is the height of the object.

For example, suppose a lamp is 30 inches from a converging lens whose focal length is 10 inches. Where will the image be found? If the lamp is 2 inches tall, what will be the height of the image? Using the lens formula:

$$\frac{1}{30} + \frac{1}{q} = \frac{1}{10}, \text{ or } \frac{1}{q} = \frac{1}{10} - \frac{1}{30} = \frac{1}{15}.$$

Inverting both sides, $q = 15$ inches. The second formula above gives $h_1 = 2 \times 15/30 = 1$ inch. Thus the image will be located 15 inches beyond the lens and will be half as tall as the object.

Exploration 10.4

You can investigate the action of a converging lens, using a simple magnifier or reading glass. The side of a small box makes a convenient stand-up screen. First measure the focal length by forming the image of a distant outdoor object, such as a tree. Move the screen back and forth until the image is sharpest and then measure the distance from the lens. Now, in a darkened room, set up a candle or a lamp at some distance from the lens greater than its focal length and locate the image by trial. Measure p and q and check the lens equation. Note also the relative sizes of object and image and compare with the ratio of their distances from the lens.

Diverging lenses do not form real images under any circumstances; they are used mainly in combination with converging lenses. If you replace the lens used in the above experiment by a diverging lens, you will get no image on the screen at any distance, but only a circle of uniform light. When parallel light strikes a diverging lens, the refracted rays spread apart from each other, all of them *seeming* to come from a point on the near side of the lens (Figure 10.16). This is a virtual focus. The distance of this point from the lens is called the focal length, just as for a converging lens. Diverging lenses always form virtual, erect images that are smaller than the object. The lens formula can be modified to take care of the case of a diverging lens.

Figure 10.16

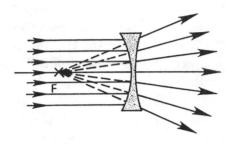

The Camera and the Human Eye

In Exploration 10.4, you put together what is basically a *camera*, except that it had no provision for keeping out unwanted light and no film for making a permanent record of the image. Also, all but the simplest photographic cameras have a combination of lenses rather than a single lens. Multiple lenses improve sharpness and other characteristics of the image. Most cameras have a provision for varying the size of the lens opening and to regulate the brightness of the image when pictures are taken under various lighting conditions. The action of the light on the film produces molecular changes in the sensitive material. Subsequent chemical treatment (development) brings out a visible image that is a negative of the original scene; it is dark where the original was light, and light where it was dark. Finally, a positive print is obtained by passing uniform light through the nega-tive onto sensitized paper, which is chemically developed to give the finished picture.

The *human eye* is optically very similar to a camera—a movie or TV camera, as a matter of fact, because the image is continually changing. An enclosure, the eyeball itself, contains a lens-shaped organ whose focal length can be shortened by muscles capable of squeezing it into a thicker shape (Figure 10.17). A watery fluid in front of the lens and a jelly behind it contribute to the refraction, producing an image on the *retina*, or sensitive rear surface. Although the image on the retina is inverted, we have learned to interpret it right side up. The retina contains millions of delicate nerve endings whose sensation is carried to the brain. The *iris* is a variable-sized diaphragm controlling the amount of light entering the eye.

Figure 10.17

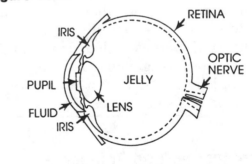

Exploration 10.5

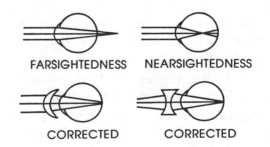

FARSIGHTEDNESS NEARSIGHTEDNESS

CORRECTED CORRECTED

Eyeglasses are probably used in greater numbers than any other optical instrument. In a normal eye, the muscles are able to compress the lens enough so that objects as close as about 10 inches away can be seen distinctly. A *nearsighted* eye is too long, front to back, to form sharp images of such nearby objects. The rays cross in front of the retina instead of on it (upper left, above). The remedy is to place a diverging lens of the proper focal length in front of the eye as shown lower left, above. In a *farsighted eye* (above, upper right) the rays strike the retina before they have had the chance to cross, and a suitable converging eyeglass lens is used to correct this difficulty (above, lower right). In an eye having the defect of *astigmatism,* the front surface of the eyeball is not curved equally in all directions like a sphere, and this produces indistinct images. Prescribing a cylindrical lens—one that is curved in one direction only—remedies this condition. If there is both astigmatism and near- or far-sightedness, the two kinds of lens needed for correction are combined in a single piece of glass.

Microscopes and Telescopes

A compound microscope is a combination of two converging lens systems—an **objective** of very short focal length and an **ocular** of moderately short focal length. Figure 10.18 shows the arrangement, each system being represented for simplicity as a single lens. The thing to be examined is brought up close to the objective, which forms a real image of it inside the tube.

This image is not caught on a screen but merely exists in space. The ocular, used as a simple magnifier, then produces an enlarged, reversed virtual image out in front of the instrument, the eye converts this into a real image out in front of the instrument, and finally the eye converts this into a real image on the retina. By special methods used by metallurgists and medical researchers, magnifications of as much as 2500 are attained. The electron microscope operating on a different principle, can attain magnifications ten or more times as great (see Chapter 16).

Figure 10.18

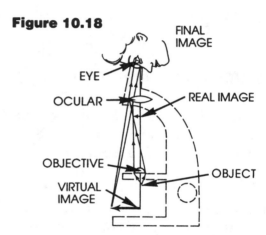

A refracting telescope, like a microscope, consists of an objective that forms a real image and an ocular for magnifying it, but here the objective has a very long focal length. Because of the optical unsteadiness of the atmosphere, magnifications of more than about 1500 to 2000 are seldom used in astronomy. Even more important to astronomers than magnification is the light-gathering power of a telescope, which determines how faint a star can be and yet be seen. This depends on the area of the objective and is one reason for making telescopes of large diameter.

To observe distances too far away for light to be visible, radiotelescopes are used to monitor electromagnetic radiation emitted by bodies far out in space.

Testing Your Knowledge

10.1 Of the following, the only object that would be visible in a perfectly dark room would be
   a. a mirror.
   b. any light-colored surface.
   c. a red hot wire.
   d. a disconnected neon tube.

10.2 The length of time it takes light to go from the sun to the earth, 93,000,000 miles distant, is about
   a. 2 min.
   b. 50 sec.
   c. 8⅓ min.
   d. 173 hr.

10.3 A man 6 ft tall stands 12 ft from a pinhole camera. When he moves 4 ft closer to the camera, his image will become
   a. 1.5 times as large as before.
   b. ⅔ as large.
   c. ⅓ as large.
   d. 1.25 as large.

10.4 Mercury is about ⅓ as far from the sun as Earth. Compared with the illumination received on Earth, that received by Mercury will be
   a. ⅓ as much.
   b. 3 times as much.
   c. 9 times as much.
   d. ⅑ as much.

10.5 In order to avoid objectionable glare when reading, one should
   a. use very low illumination.
   b. avoid anything printed on rough paper.
   c. have the light coming from over one shoulder.
   d. use a bare, unshaded lamp.

10.6 When a man walks directly toward a vertical mirror at a speed of 4 ft/sec he
   a. approaches his image at the rate of 4 ft/sec.
   b. recedes from his image at the rate of 4 ft/sec.
   c. approaches his image at the rate of 8 ft/sec.
   d. stays a constant distance from his image.

10.7 In order just to see himself from head to foot when standing before a vertical plane mirror, a man 6 ft tall must have a mirror

   a. 3 ft high.

   b. 3 × 3 ft in size.

   c. at least 3 ft wide but of any arbitrary height.

   d. 6 ft high.

10.8 What, if any, are the dimensions of n, the index of refraction of a substance?

10.9 Concave mirrors perform the same functions as converging lenses, convex mirrors the same ones as diverging lenses. Explain.

10.10 What are bifocal eyeglasses?

10.11 A camera has an objective of focal length 10 inches. Use the lens formula to find how far from this camera a person must stand in order to be in sharp focus when the objective is racked out to a distance of 10.5 inches from the film.

10.12 The head of a common pin is 1/14 inches across. If all of it could be seen at one time in a microscope under a magnification of 2500, how big across would it appear to be?

10.13 The objective mirror of the Mt. Palomar telescope has a diameter of 200 inches, while the pupil of the eye has a diameter of 0.2 inches. Since the amount of light that each receives is proportional to the area of its circle, prove that this telescope has about a million times the light-gathering power of the eye.

CHAPTER 11

Wave Optics

KEY TERMS FOR THIS CHAPTER

spectrum

primary colors

ultraviolet

infrared

X rays

diffraction

interference of light

polarization of light

Think of a sunset . . . a beautiful painting . . . the car you'd like to own . . . a jewel or a flower or an article of clothing. Chances are, you are thinking first of its color. Color makes us want to buy or eat or wear something; it can make us hotter or colder, even happy or sad. In this chapter, we talk about the phenomenon we call color, how colors can be mixed to produce other colors, and about other aspects of light that have to do with its wavelength.

The Spectrum

Ancient peoples were aware of the brilliant hues produced when sunlight passes through transparent gems and crystals. Isaac Newton, while still a university student, performed a simple experiment that revealed the true character of color. Holding a triangular glass prism in the path of a narrow beam of sunlight, he found that the rays were fanned out into a band of

The spectrum is the sequence of colors that results when light is dispersed through a prism. You can see this order in a rainbow.

color after passing through. The sequence was the same as the one seen in the rainbow, with red at one end merging gradually into orange, then yellow, green, blue and violet. (Sometimes an additional range of color, indigo, is designated between blue and violet.)

Newton suspected that these shades are all present to begin with in the colorless beam of sunlight—what we call white light—and that in passing through the prism, the various colors are refracted by very slightly different amounts. Once the colors are spread out this way, the eye is able to recognize each for what it is. The spreading process is called **dispersion**, and the color sequence that results is called the **spectrum**.

Newton tested his idea about the nature of white light by placing a second prism behind the first, but in a reversed position. He found that the rays were reunited into a colorless path of white light. As a further experiment, he cast a spectrum on a card, cut a small hole in the card at one point, and placed another prism behind the opening to see whether the color coming through the hole could be further broken up by the second prism. This did not happen. For instance, green light coming through the opening was bent aside by the second prism, but the transmitted light was still green.

What Newton established, then, is that *white light consists of a mixture of colors.*

(Black light is not a color but the absence of color.) The different colors are refracted by slightly different amounts in glass, which means that the index of refraction is really slightly different for each color. Assuming that light consists of waves, what we perceive as color is really wavelength, and that the waves representing various colors travel with different speeds in matter. The difference is not very big: In ordinary glass, red light travels only about 1 percent faster than violet. In a vacuum, all colors travel with the same speed. At the extreme red end of the spectrum the wavelength is about 1/30,000 of an inch (0.00007 cm), decreasing gradually through the sequence of colors to about 1/60,000 of an inch (0.00004 cm) at the extreme violet end.

Exploration 11.1

Repeat some of Newton's tests. If you have no optical prism (a small one can be bought very cheaply from dealers in salvage materials), use a crystal glass pendant, or make a water prism by holding a rectangular glass dish, partly filled with water, at a slant. Let sunlight hit the prism and catch the spectrum on a white card. Shield the card from the direct sunlight. Now put the prism, edge up, just beyond the lens, and the white slit image will give way to a colored

White light is a mixture of colors; black is the absence of color.

spectrum on the screen. Recombine the colors into a white spot of light by using another lens or a concave mirror to bring them together.

Since different wavelengths are refracted by slightly different amounts, some separation of color occurs whenever light passes through any transparent material. The rainbow, one of the most commonly observed examples of dispersion, is formed when sunlight passes through myriad droplets of water suspended in the air after a shower (Figure 11.1). The rays are refracted on entering a drop, reflected from the back surface, and refracted again on coming out. But while going through, the light is also dispersed, so that the several colors come through at slightly different angles. The total effect produced by all the drops is the observed rainbow.

Figure 11.1

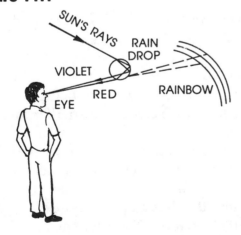

Mixing Colors and Pigments

In saying that the spectrum is made up of the colors red, orange, yellow, green,

blue, and violet, we make these divisions merely for convenience. The whole spectrum really consists of an endless number of colors (wavelengths). Experts can distinguish thousands of shades. The eye also recognizes colors that are not found in the spectrum, but which are effects produced by mixing various spectrum colors. Purple, for instance, is a mixture of blue and red; pink is a mixture of red and white; browns are produced by mixing red and yellow.

Virtually any recognizable color can be produced by mixing different proportions of the three colors red, green, and blue-violet, which are called the **primary colors.** For example, if three such spots of light are thrown on a screen, the place where all three overlap will appear white; other regions, where only two overlap, give different colors as shown in Figure 11.2.

Figure 11.2

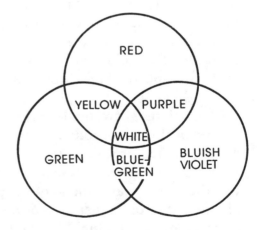

The primary colors are red, green, and blue-violet.

Exploration 11.2

Use a red and a green felt-tipped pen or watercolors to mark a white sheet of paper. Examine the marks first under red and then under green light. (Hold a sheet of colored cellophane in front of either the lamp or your eye.) Explain what you observe.

What happens when various colors, or wavelengths, of light are mixed and reflected to your eye is not to be confused with what occurs when different paints or pigments are mixed. For example, when yellow and blue-violet *light* are mixed as in Figure 10.2, the result is white. However, when yellow and blue-violet *paints* are mixed, the outcome is green paint. Paints or pigments absorb (subtract out) certain wavelengths from white light and reflect the rest. In the case just mentioned, the yellow paint absorbed blue and violet and the blue absorbed red and yellow, leaving green as the only color that the mixture could reflect to your eye.

An apple looks red in ordinary daylight because its skin contains pigments that can absorb the other colors. The same apple seen in red light appears pale and whitish because it reflects nearly all the light falling on it. In green light, however, it looks black, because it is not able to

---

EVERYDAY PHYSICS: 11

The photographs reproduced in newspapers and magazines are composed of many dots—just like the image on your TV screen. The dots are easy to see in a newspaper, because the paper is coarse and the dots—even in the "solid black" areas—are really quite far apart. But even the photographs in a fine "coffee-table" art book are reproduced in this same way, by being rephotographed (shot) through a screen. The finer the screen, the smaller and closer the dots.

Color pictures are made of dots, too. Any printed picture, even one of the most colorful, is reproduced using only three or four pigments plus black. In a complicated process, one printing plate is made for each color; when the paper is run through these plates in succession, the dots that are close together appear as a blend of the pigments they contain. Your eye then "mixes" the colors just as you might mix them on a palette. Making color printing plates is a highly skilled process; computers have made it quicker and easier because a picture can be scanned and the wavelengths of the colors digitized (converted into numerical signals) to make the printing plates.

Objects appear to change color in different lights because their pigments reflect or absorb different wavelengths.

---

reflect much of the incoming light (remember that black is merely the absence of light). We see from all this that the apparent color of an object depends very much on the quality of the light by which it is seen. Recent experiments by Edmund Land and others have shown that the brain compensates in some way for minor variations in light to keep colors apparently constant. Still, the effect of light on color perception must be kept in mind in any practical operation that involves the matching of colors, for example in the textile industry.

The chemical processes that go on in plants require red light. While a plant is alive, the red part of sunlight is absorbed by a substance called chlorophyl, and so the remaining wavelengths—largely green—are reflected, giving the plant its characteristic color.

Measuring and Analyzing Light and Color

The simple arrangement Newton used to produce a spectrum can be improved in several ways: In place of a round hole for admitting the light, a narrow slit perhaps less than a hundredth of an inch wide is used. This greatly reduces overlapping of colors in the final spectrum. Lenses are

placed before and after the prism to improve the definition, and the spectrum is magnified. The complete instrument is called a *spectroscope*. If a permanent record is wanted, the magnifying lens is omitted and a photographic film is put in the focal plane of the second lens. In this case the instrument is called a *spectrograph*.

When the radiations from various substances are examined and analyzed by a spectroscope, we find that sunlight, the light from a filament lamp, a pot of molten iron, or a candle flame all give the rainbowlike band of color already described. This is called a *continuous spectrum;* all wavelengths are present and there are no gaps from extreme red to extreme violet.

The situation is different when gases or vapors are made to give off light. Then what is seen in the spectroscope is a number of sharp, bright lines against a dark background. The dark areas are evidence of quantum leaps of energy. Each line is an image of the slit at a given wavelength of light, and so we conclude that a gas can give out only a definite set of wavelengths. Figure 11.3 shows the line patterns for hydrogen and oxygen. The pattern of lines is different for each chemical element present in the light source, and so once we know the patterns, we can identify the chemical elements present in any source. This process, which is called *gas chro-*

By analyzing the wavelengths of light given off by gases, we can identify the chemical contents of the sun and other stars.

---

matography, enables us to find out what elements even the sun and stars contain. We can even find out the proportions of each substance by measuring the brightness of the lines.

Figure 11.3

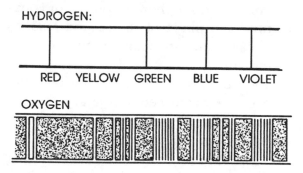

These techniques of *spectrochemical analysis* are taking the place of regular chemical methods in many industries, in medicine, and in crime detection.

Electromagnetic Waves: A Preview

The human eye can see only light whose wavelength lies between about 1/30,000 and 1/60,000 of an inch, but there are other waves of a similar nature—some shorter, some longer—that differ from visible light only in the ways by which they can be produced and detected. The waves that fall beyond the violet end of the visible spectrum make up the **ultraviolet** region; the ones beyond the red end are called the **infrared.** Each is present to some extent in sunlight, but the shorter ultraviolet waves are much stronger in the light of a mercury vapor lamp (sunlamp), for instance, and infrared is given off generously by a hot stove or electrical heating coil.

X rays are still shorter than the ultraviolet, about one thousandth the wavelength of visible light. X rays are produced by a Coolidge tube in which highly accelerated electrons are aimed at a tungsten filament heated to high temperature. The outstanding characteristic of X rays is their ability to penetrate through most materials. Besides their use in medicine, they have uses such as revealing flaws in metal parts and alterations of paintings and documents. Even shorter and more penetrating than X rays are the *gamma rays* given off in atomic processes. Their uses are similar to those of X rays.

On the long-wave side of the infrared are radiations that are produced by electric circuits in which alternating currents flow. These *electric waves* include the ones used in radio and television. All these types, from the shortest gamma rays to the longest electric waves, are of the same essential character—they are *electromagnetic waves*. All behave like light, and can be reflected, refracted, diffused, and dispersed. All travel through empty space with the same speed. Electromagnetic waves are discussed at length in Chapters 12 through 16.

Diffraction

Recall from Chapter 10 that light travels in straight lines while passing through any uniform region. Of course, if light strikes a mirror or if the density of the material through which it passes is not the

---

The bending of waves when they pass near the edge of an obstacle or through small openings is called diffraction.

---

same at all points, it will change its direction.

We discussed these possibilities in connection with reflection and refraction. But even in uniform space, light and other forms of wave motion are found to bend somewhat if they graze the edge of a barrier or go through very small openings. In fact, Huygens first got the idea that light consists of waves when he noticed that sea waves spread out sidewise when coming through the opening in a breakwater. Certainly we are all familiar with the fact that sound waves bend around corners very considerably—you do not have to be in line with an open window to hear sounds from outdoors. This bending of waves when they pass near the edge of an obstacle or through small openings is called **diffraction**. The fact that it can be observed for light, under proper conditions, is strong evidence in favor of the wave theory.

Exploration 11.3

Use a tray of water, as you did in the set-up for Exploration 9.1, to produce diffraction of water waves. Line up two boards or other barriers across the tray, with an opening about an inch wide between them. Observe the increased spreading of ripples after they pass through the gap, and notice that if you make the aperture smaller, the bending becomes more pronounced.

When white light is reflected from the edge of a phonograph record (see Exploration 12.1), you see an iridescent rainbow play of colors because the different wavelengths of light are diffracted by different amounts when reflected by the regularly spaced ridges of the record surface. In fact, a surface covered by fine, evenly spaced channels or ridges can be used as a substitute for the prism in a spectroscope. Such *diffraction gratings* are made by special machines that rule extremely fine scratches on metal or glass plates by means of a diamond point. A good grating of this kind may have 15,000 or more rulings to the inch and is capable of giving much greater dispersion than any prism.

As fine as such optical gratings are, they are too coarse for producing diffraction of the very much shorter X rays. But crystals of certain minerals can serve as natural gratings for this purpose. The regular spacing of the atoms in a crystal is just of the right order of size for diffracting X rays and thus can serve to measure their wavelengths. Then, using X rays of known wavelength, the exact arrangement of the atoms in other crystals can often be worked out.

Exploration 11.4

Look at a distant lamp through a handkerchief or through the fabric of an umbrella. In place of a single spot of light you will see a square arrangement of fainter, smeared-out colored images due to diffraction of light by the two sets of threads of the cloth—really two diffraction gratings at right angles to each other.

Interference of Light

The beautiful display of colors you see on a soap bubble or on a spot of oil on a wet pavement is not due to pigments or to diffraction but to the **interference of light.** It depends on the fact that two sets of waves arriving at the same place will add up their effects if they get there *in step,* but will cancel each other if they get there *out of step.* The English scientist Thomas Young discovered the interference principle around 1800 and saw that it furnishes further confirmation that light may be a wave motion.

In all examples of interference, light from a single source is, in effect, reflected from two surfaces that are very close together. When you look at a soap film, for example, part of the light that comes to you has been reflected from the front surface, part from the back (Figure 11.4). If the two reflected rays of a given wave-

length come back exactly out of step with each other, they cancel, and no light of this color reaches your eye from this point of the film. Whether interference of light happens depends on both the wavelength of the light and the thickness of the film at that place. Since white light consists of many different wavelengths and since a soap film is not of uniform thickness, the result is that you see a variety of colors, each one being white light *minus* the particular wavelength that is cut out by interference. For example, where green is destroyed, the film will have a reddish hue.

Figure 11.4

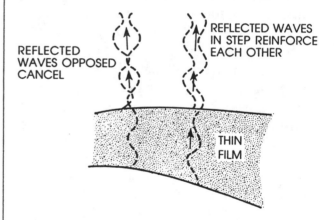

A useful application of interference is in nonreflecting coatings for glass. The surface is covered with a chemical film of just the right thickness to kill off most of the light that would ordinarily be reflected and cause glare. When applied to a camera objective this improves the quality and brightness of the image by cutting out reflections from the various lens surfaces. An arrangement of the type you made in

Exploration 11.5 evidently can magnify very small motions. The various practical forms of this device are called interferometers. They are widely used in science and in precision industries. With them it is possible to measure distances of the order of a millionth of an inch.

Exploration 11.5

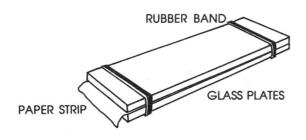

RUBBER BAND

GLASS PLATES

PAPER STRIP

Make your own interferometer. Get two rectangular strips of clear glass several inches long and place one on top of the other. Fasten their ends together with rubber bands and separate the plates slightly at one end by inserting a strip of paper. Now take a piece of paper toweling that was previously soaked in strong salt water and dried, and hold it in a gas flame in a darkened room. This produces a yellow light of a single color, which gives better contrast than white light. Look at the reflection of the flame in the pair of plates and you will see that they are crossed by evenly spaced dark stripes as a result of interference of light reflected from the front and back of the "wedge of air" between the plates. Squeeze the pair more tightly together at one end and the stripes will move some distance along the plates.

Polarization

Refraction, interference, and diffraction all serve to substantiate the wave theory of light. But what *kind* of waves are light waves—longitudinal, like sound? Transverse, like waves in a rope? Or a mixture of the two, like sea waves? The answer is that light waves are transverse. The proof comes from a study of what is called the **polarization of light.**

Consider what happens when a beam of ordinary light passes through certain crystals in which the arrangement of the atoms give the crystal a sort of ribbed structure equivalent to a large number of parallel slots. If two such crystals are placed one behind the other, light will get through both when their "slots" are parallel, but will be completely cut off if the slots are crossed. The only assumption that will explain this observation is that the vibrations of ordinary light are in all possible directions in a plane perpendicular to the ray. A single crystal will then hold back all the vibrations except the one that is lined up with its own grain. A beam of light whose vibrations are thus confined to one direction is said to be *plane-polarized*. This polarized beam gets through the second crystal when the two sets of slots are parallel but will be gradually cut off as the second crystal is rotated to the perpendicular position.

A manufactured polarizing sheet material called Polaroid has now replaced natural crystals for most applications of polarization. For example, most sun glare is caused by reflection of rays in the vertical direction. Polaroid sunglasses filter the light so that only the horizontal rays pass through, thus eliminating the glare.

Polarized light can be used to find just how the stresses are distributed in machine parts. A model of the part is made out of plastic and subjected to the kind of stress the original would get in actual use. When the object is viewed by polarized light, colored bands appear which reveal the exact stress pattern in the piece.

Figure 11.5

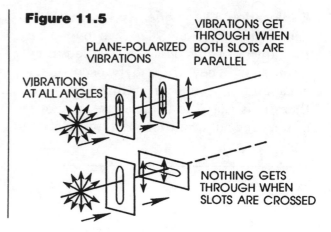

Testing Your Knowledge

11.1 From Figure 11.1 it is evident that the color of the outer edge of the rainbow will be
a. purple.
b. red.
c. violet.
d. yellow.

11.2 The effect produced in the eye by a mixture of purple and yellow light will be
a. white.
b. blue-green.
c. blue-violet.
d. green.

11.3 A tailor, wishing to match thread for a blue overcoat, should do so
a. under a yellow light.
b. under a blue light.
c. near a window.
d. in semi-darkness.

11.4 A certain piece of paper looks red under red light but appears black under blue light. When seen in daylight, it may appear
a. colorless.
b. blue.
c. white.
d. red.

11.5 When seen through a piece of red glass, the leaves of a plant
    a. appear almost black.   c. are seen in their natural color.
    b. become nearly invisible.   d. take on a bluish blue.

11.6 For light traveling in a vacuum, the wave equation can be written $c =$ frequency $\times$ wavelength. What is the frequency of vibration of orange light of wavelength 0.00006 centimeters?

11.7 Describe the kind of spectrum you would expect to get from (a) moonlight; (b) a luminous tube containing neon gas; (c) a candle flame whose light is due mainly to glowing solid particles of carbon.

11.8 The Doppler effect is observed for light. Light waves from a star approaching the earth are shortened. Would the lines of its spectrum be displaced toward the red end of the spectrum or toward the violet?

11.9 When you look at the moon through a window screen you see a cross of light through the moon's image. Explain.

11.10 An inventor once suggested distilling soap solution in order to get out the colored dyes seen on soap bubbles. Is this a sound scientific idea?

PART THREE

ELECTRICITY AND MAGNETISM

CHAPTER 12

Electric Charge

KEY TERMS FOR THIS CHAPTER

static electricity

electric current

insulator (nonconductor)

conductor

grounding

electrostatic induction

electric field

coulomb

joule

potential difference

capacitance

Electrical power, harnessed to our use, is so much a part of the everyday scene that it's hard to imagine life without it. We cook with it, compute with it, use it to hear music, to record images, and even to explore stars so far away we cannot see their light. But modern and "high-tech" as these uses are, our name for the force behind them goes back 25 centuries to ancient Greece, where a fundamental observation was recorded. Amber, if rubbed with a cloth, could attract bits of straw. Later it was discovered that other materials besides amber have this characteristic, which is called electrification. The word is derived from the Greek word for amber.

Static Electricity and Electric Current

Run a comb through your hair in dry weather, and you'll hear a crackling noise. Walk over a rug or slide across a car seat and touch metal, and you get a slight electric shock. Pet your cat in the dark and see sparks. All these experiences demonstrate that any two substances rubbed together under suitable conditions become electrically charged. An electric charge resting on an object such as your comb is called **static electricity.** A charge in motion is an **electric current.** It is electric current that we use in most electrical appliances.

A resting electric charge is called static electricity; a charge in motion is called an electric current.

You can electrify a stick of hard rubber or sealing wax by rubbing it with a piece of fur or flannel. You can also electrify a glass rod or tube by rubbing it with a silk scarf. In either case, the attraction of bits of paper shows a charge to be present, but in a very definite way the charge on the hard rubber and the charge on the glass are of an opposite nature: Two electrified pieces of hard rubber will be found to repel each other; but an electrified piece of hard rubber rod and an electrified piece of glass will attract each other (Figure 12.1). Such experiments show that charges are of two kinds, and that *unlike kinds attract each other, while like kinds repel*. It was Benjamin Franklin who gave the names "positive" and "negative" to the two sorts of charge. The kind found on the glass rod he called positive (+), that on the hard rubber, negative (−).

Figure 12.1

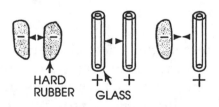

HARD RUBBER   GLASS

Atoms and Electric Charge

The charging of bodies is now understood in terms of the structure of the atoms composing the body. We will discuss these as facts here, but the data underlying that understanding are discussed in later chapters.

Every atom consists of a compact central nucleus carrying a positive charge, around which are distributed a number of electrons, each negatively charged. The nucleus owes its positive charge to the fact that it contains particles called protons. All protons are alike, and all carry the same amount of + charge. All electrons are alike, and each carries a − charge equal in amount to the + charge of a proton.

In its normal condition, the atom is electrically neutral—there are just as many electrons outside the nucleus as there are protons inside it (Figure 12.2). However, certain atoms are able to hold *temporarily* more than their normal number of electrons, so when two different materials such as glass and silk are put into good contact by rubbing, the glass gives up some electrons to the silk. Thus the silk now has a negative charge, while the glass has a *deficiency* of negative charge, which is considered to be a positive charge.

There are two kinds of electric charges, positive (+) and negative (−); like charges repel, unlike charges attract.

Materials that allow electrons to pass freely are called conductors; those that do not are called nonconductors or insulators.

Charges are not "produced" by rubbing, they are merely separated out. It is usually the electrons that move from one place to another, the nuclei of the atoms remaining fixed in place.

Figure 12.2

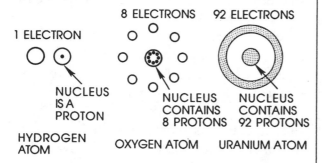

1 ELECTRON

NUCLEUS IS A PROTON

HYDROGEN ATOM

8 ELECTRONS

NUCLEUS CONTAINS 8 PROTONS

OXYGEN ATOM

92 ELECTRONS

NUCLEUS CONTAINS 92 PROTONS

URANIUM ATOM

Conductors and Insulators

A charged body, if it is to keep its charge, must be supported by something that does not allow electrons to pass freely along it. Suitable materials, such as glass, hard rubber, sulfur, porcelain, are called **insulators** or **nonconductors**. On the other hand, many substances, especially metals, allow charge carriers to pass easily and are called **conductors** of electricity.

If a charged rod is touched to an insulated metal object, some of the charge will pass from one to the other, and the metal body is said to receive a charge by conduction. This charge will spread over the entire surface of the conductor. If the charged object is now connected to a very large body (such as the earth) by means of a wire, it is said to be **grounded** and loses its charge. If negative to begin with,

its excess electrons flow off through the ground wire; if originally positive, the required number of electrons come up to it from the ground. The earth acts merely as a very large storehouse of charge of either kind.

Because a charge always goes entirely to the outside of a conductor, a sheet-metal box or even a wire cage that is grounded will act as an electrical shield, cutting off the effects of charges that may exist on the outside. Radio sets and other electrical apparatus can thus be shielded against external disturbances. If a charge is put on a pointed conductor, most of the charge piles up at the point. If the point is quite sharp, the charge may actually leak off to the surroundings (Figure 12.3).

Figure 12.3

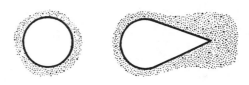

The discharging effect of points is used in the electrical smoke and dust precipitator. The particles become charged by the electricity streaming from the points (Figure 12.4), are thus driven away, and collect on the grounded plate. Not only is pollution of the air avoided, but the reclaiming of valuable materials from the dust may make the whole operation quite profitable.

EVERYDAY PHYSICS: 12

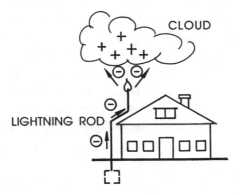

Franklin's famous kite experiment identified lightning as a discharge of electricity—an electric spark. Following this up, he devised the lightning rod. A lightning rod is merely a pointed conductor placed at the top of a structure and connected by a heavy wire to a metal plate buried in moist earth. Suppose a positively charged cloud passes over a house. Its attraction makes electrons flow up to the rod from the earth. On reaching the pointed rod, the electrons leak off and quietly neutralize the charge of the cloud before it can cause damage by suddenly finding a path to earth through the structure itself.

Figure 12.4

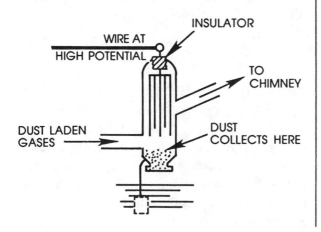

Electrostatic Induction

Charges can be induced in a neutral body by bringing it near a charged one. In Figure 12.5, the insulated, uncharged object is represented as having a uniform mixture of + and − charges all through it. When the + rod is brought near one end, some of the electrons are attracted toward that end. Now, touching the body with a finger provides a ground connection through which additional electrons may come onto the body in response to the attraction of the + charge on the rod. If the ground connection is now broken, the body will be left with an excess of electrons, and when the rod is finally removed they will redistribute themselves more uniformly. Although it is never actually touched by a charged object, the body acquires a charge by induction. In a similar way, a negatively charged rod can be used to give an object a + charge.

Figure 12.5

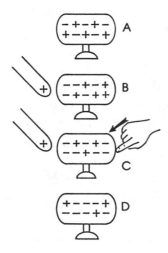

Exploration 12.1

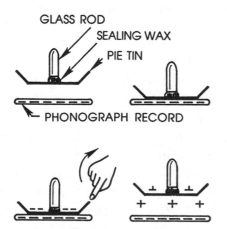

GLASS ROD

SEALING WAX

PIE TIN

PHONOGRAPH RECORD

Make an electrophorus (a device for generating electrostatic charges). Fix a glass rod or small bottle to the inside of a pie tin with sealing wax to make an insulating handle. Place an old phonograph record on the table and rub it briskly all over with a piece of fur or wool. Set the pie tin down on the record and touch the tin momentarily with your finger to ground it. When lifted away, the tin will be electrified enough to enable the charge on it to jump as much as a quarter of an inch through the air to your knuckle. Without rubbing the record again, you can get a fresh charge on the disk over and over before the original charge on the record has leaked away.

The diagram shows the stages in the charging process. The work you do each time the disk is lifted away from the record in opposition to the attraction between the charges on the two pays for the seemingly unlimited amount of electrical energy that is produced. Continuous-operating electrostatic generators based on a similar principle produce large amounts of charge at millions of volts. They are used in studying the effect of lightning on power lines and for "atom smashing" experiments.

Fields of Electricity

The space in the neighborhood of charged objects is called an **electric field.** Lines of force are used to map out electric fields. Each line gives the direction of the resultant force on a small + charge placed at the point in question. The lines are

---

The force between two charged bodies is directly proportional to the amounts of charge and inversely proportional to the square of their distance apart.

---

thought of as originating on + charges and ending on − charges. We will see (in Chapter 14, Magnetism) that the magnetic field between opposite poles is exactly that between opposite electrical charges. The force between two charged bodies is found to be directly proportional to the amounts of charge and inversely proportional to the square of their distance apart. The amount of charge on an object can be measured by bringing a standard charge up to a certain distance and measuring the force that one exerts on the other. The practical unit of quantity of charge is called one **coulomb,** after the French experimenter of that name. One coulomb is about the amount of charge that flows through a 100-watt filament lamp every second, yet it is equivalent to 6.3 billion billion electrons.

Potential and Capacitance

If a steel ball is placed on a hill, it will roll down, and we can say that it does so because there is a resultant downhill force acting on the ball. But we can also describe what happens in another way by saying that the ball will move from a position of higher gravitational potential en-

ergy to one of lower. Or, in discussing two connected water tanks, we can say that water will have a tendency to flow from the tank with the higher water level to the one with the lower level because at the connecting point the pressure is greater on the side where there is more water. When the valve is opened, the flow will take place and will continue so long as any pressure difference exists. In a corresponding electrical case we say that a charge will have a tendency to move from one place to another if an electrical **potential difference** (PD) exists between the two places. For instance, if two insulated metal balls are at different electrical potential, charge will flow from the higher to the lower when they are joined by a conducting wire. If the potential difference is high enough to begin with, the insulating ability of the air between them may be insufficient, and a spark will pass from one to the other.

The ability of a conductor to take on more charge as its potential is raised is measured by its electrical **capacitance.** The capacitance of a conductor may be increased greatly by putting a grounded conductor close to it. A "sandwich" consisting of two flat metal plates separated by a thin sheet of insulating material such as air, glass, mica, or waxed paper con-

---

The practical unit of quantity of electric charge is the coulomb; 1 coulomb is equal to 6.3 billion billion electrons.

An electric charge will tend to move from an area or object having a higher charge to an area or object having a lower charge.

stitutes an electrical capacitor. The charge-storing ability may be increased by using many layers, with the alternate plates connected together (see Figure 12.6). Capacitors are indispensable parts of radio, telephone, and TV circuits, and of many other electronic devices.

Figure 12.6

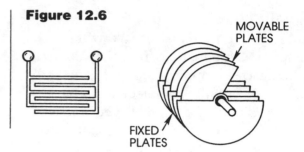

MOVABLE PLATES

FIXED PLATES

Testing Your Knowledge

12.1 Using the idea of electrostatic induction, make sketches to explain why any neutral object will be attracted by a charged rod carrying either + or − electricity.

12.2 A small cork ball hanging by a silk thread is attracted when a charged glass rod is brought near, but if the ball moves over and touches the rod, it will bound away immediately afterward. Explain.

12.3 During a thunderstorm, violent rising currents of air carry drops of water upward within the clouds. Can you explain how these drops become charged?

12.4 Friction between tires and road sometimes causes a considerable charge to accumulate on a car, and the potential of the car body may reach several thousand volts. What is the function of the flexible strap that some motorists hang from the axle of their car?

12.5 An insulated metal ball has an excess of 3 billion electrons on it, and an identical metal ball has a deficiency of 4 billion. (a) Do they attract or repel each other? (b) If the two are touched together and then separated, what will be the kind and amount of charge on each one and the nature of the force between them?

CHAPTER 13

Electricity

KEY TERMS FOR THIS CHAPTER

ampere

voltaic cell

volt

dry cell

lead storage cell

electrical circuit

electrical resistance

Ohm's law

series circuit

parallel circuit

watt

kilowatt-hour

fuse

superconductors

An electric current is an electric charge in motion. In a solid conductor such as a wire, the current consists of a swarm of moving electrons, while in certain liquids and in gases the carriers may include positively and negatively charged atoms. In addition, a beam of electrons or charged atoms may be made to go through a vacuum, no conductor being involved at all. Such a beam amounts to a current just as much as one in a wire. In this chapter, you will find a description of the basic facts concerning the flow of electricity in circuits consisting of solid and liquid conductors.

Measuring Electric Current

Electric current flows through wire conductors much as water flows through pipes. The flow of water past any point in the system can be measured by units and time: gallons per second, cubic feet per hour, and so forth. The strength (flow) of electric current, called simply "current" from now on, is measured by the amount of charge passing a given point per unit of time, stated in amperes. This is expressed mathematically as:

$$I \text{ (current strength)} = \frac{Q}{t} \begin{array}{l} \text{(quantity of charge)} \\ \text{(divided by)} \\ \text{(time in which it passed)} \end{array}$$

The **ampere,** named for the French scientist and mathematician André Marie Ampère, is a rate of flow of one coulomb of charge per second, which means 6.3 billion billion electrons per second. In a metal, this vast number of electrons is closely crowded together, so that their movement in a current of moderate strength amounts to a slow drift. The reason a light goes on instantly when you flick a switch is not that electrons race around to it at high speed but that the conductors are always "filled" with electrons, just as a pipe is always filled with water that gushes out when you open the faucet.

A water system consisting of a series of pipes joined to a circulating pump corresponds to a simple electrical circuit made up of a series of wires connected to a battery (Figure 13.1). The purpose of the pump is to maintain the pressure difference between its inlet and outlet in order to keep the water circulating. The function of the battery is to maintain an electrical potential difference (see Chapter 12) between its two terminals. This potential difference keeps the current going in the circuit.

Figure 13.1

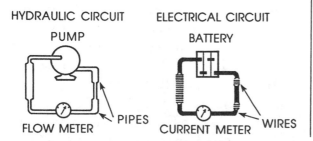

HYDRAULIC CIRCUIT   ELECTRICAL CIRCUIT
PUMP   BATTERY

FLOW METER   PIPES   CURRENT METER   WIRES

How Voltaic Cells Work

How does a battery accomplish this effect? In a famous series of eighteenth-century experiments, Italian biologist Luigi Galvani found that the muscle of a frog's leg would twitch when it was touched at the same time by two metals, such as brass and iron. Galvani believed the movement was due to some kind of "animal electricity." However, Count Alessandro Volta, an Italian physicist, showed that similar effects could be produced without using animal tissue at all.

Volta built the first storage battery. He stacked alternate zinc and copper disks separated by pieces of leather soaked in salt solution. From this he obtained the same kinds of action as from a charged capacitor, except that the operation could be repeated again and again. A single unit of such a battery is called a **voltaic cell.** Even though such cells were in use, chemists could not explain how they worked until long after the time of Galvani and Volta. The typical chemical cell represented in Figure 13.2 is made by placing a rod of zinc (chemical symbol Zn) and a rod of copper (Cu) in a solution of hydrochloric acid (HCl).

When hydrochloric acid molecules dissolve in water they break apart, or dissociate, into two pieces one of which is a chlorine atom with an extra electron attached to it. This is called a chlorine ion. The attached electron is indicated by writing a minus sign on its symbol: Cl^-. This electron was obtained from the hydrogen

Storage batteries maintain potential difference in an electrical circuit the way a pump maintains a pressure differential in a water system.

atom which, having had this negative charge taken away from it, has now become the positive hydrogen ion H⁺. In a similar way, many other chemical substances dissociate in solution to form ions.

Figure 13.2

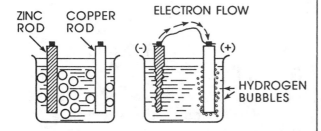

When the zinc rod is put into the liquid, the zinc atoms have a strong tendency to detach themselves from it. Each Zn atom comes off as a doubly charged ion, Z^{++} Every time this happens, a pair of electrons is left behind on the rod. Soon no more Zn ions come off, because the negative charge of the rod creates a back attraction. The accumulating Zn ions repel the H⁺ ions, making them collect near the copper rod. The copper does not dissolve to any extent, and nothing further happens until the outside circuit is completed by connecting a wire between the two rods.

When the outside circuit is completed, electrons that have piled up on the zinc rod flow over this wire to the copper rod, where they neutralize the positive charges carried there by the H ions. Having given up their charge, the H ions are again ordinary atoms of hydrogen, and hydrogen gas begins to bubble out of the liquid at the surface of the copper rod. The action goes on until the zinc rod is completely used up.

How Batteries Work

You have probably noticed that batteries are rated in volts. **A volt** is a practical measure of potential difference; it is still a standard measurement in the United States. It is equal to one joule per coulomb. A voltaic cell can maintain a potential difference of about 1.5 volts between its terminals if a very small current is being drawn from it. A battery can be made up by connecting a number of voltaic cells in a series. The potential difference will equal the number of batteries times 1.5. The most widely used voltaic cell is the **dry cell,** used in flashlights, cassette players, toys, and "laptop" computers. The construction of a dry cell is shown in cross-section in Figure 13.3.

Figure 13.3

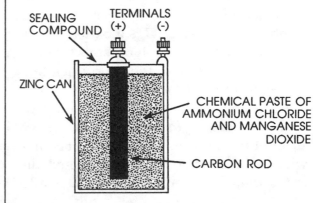

The metals that best conduct electricity are silver, copper, and gold, although aluminum is also used in electrical wiring.

A **lead storage cell** consists of a lead plate and one containing lead peroxide placed in a sulfuric acid solution. The action is similar to that of a voltaic cell, except that the plates do not dissolve but can be restored to their original state by passing a "charging" current through the cell in the opposite direction. The cell may be used repeatedly. The usual automobile storage battery is made up of such cells.

How a Simple Electrical Circuit Works

A simple electrical circuit consisting of a source of potential difference (PD) and a series of conductors was diagramed in Figure 13.1. A useful addition is a switch for opening and closing the circuit. If we want to know the magnitude of the current and the PD between any two points, suitable measuring instruments called, respectively, ammeters and voltmeters may be used.

Figure 13.4 shows an electrician's diagram of such a circuit, using the standard symbol for each part. *B* is a battery made up of three cells. The long stroke represents the + terminal, the short thick line the − terminal of each cell. The zigzag line *R* is any conductor through which we wish the current to go, while the heavy straight lines represent heavy connecting wires. The ammeter *A* is connected directly *into the circuit* at any point, while the voltmeter *V* is *in a side circuit*, its terminals being connected to the two points whose PD we wish to know—in this case,

the ends of the unit *R*. The current that passes through *V* is negligible compared to the current in the main circuit. When *K* is closed, a steady current flows in the circuit and the meters take on steady readings.

Electrical Resistance and Ohm's Law

What determines the strength of the current that flows in the circuit? Early in the nineteenth century, German scientist G. S. Ohm conducted experiments to answer this question. By connecting pieces of wire of various lengths, cross-sections and materials in place of *R* in a circuit like the one in the diagram, Ohm found that the current is directly proportional to the cross-section area of the wire and inversely proportional to its length, and also depends on the kind of metal of which the wire is made. The best conductors are found to be silver, copper, and gold.

Figure 13.4

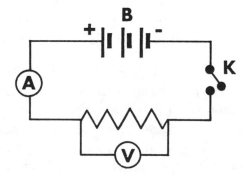

Ohm's law states that for a given wire, the strength of the current is proportional to the potential difference between the ends of the wire.

A wire offers resistance to current flow just as a pipe offers resistance to water flow. Ohm found that for a given wire in a circuit, the current is proportional to the potential difference between the ends of the wire, or:

$$I \text{ (current strength)} = \frac{V}{r} \quad \begin{array}{l}\text{(applied} \\ \text{potential} \\ \text{difference)} \\ \textit{divided by} \\ \text{(resistance of} \\ \text{conductor)}\end{array}$$

This is the famous **Ohm's law** of current electricity.

The practical unit of I is the ampere; the practical unit of V is the volt. The corresponding unit for resistance (R) is 1 **ohm.** A 1-ohm resistor is one that allows a current of one ampere (amp) to flow when a potential difference (PD) of 1 volt is applied to its ends. For example, the resistance of the hot filament in a 60-watt bulb is over 200 ohms. The total resistance of a simple circuit like the one in Figure 13.4 may be only a few hundredths of an ohm.

The most direct way to find the resistance of a conductor is to put it in a simple circuit, measure the potential difference across it and the current through it by suitable meters, and then compute the resistance using Ohm's law. For example, what is the strength of a current through a filament lamp if the resistance of the filament is 220 ohms and it is used in a 110-volt line? The applied potential difference can be taken to be 110 volts, so

$$I = {}^{110}\!/_{220} = 0.5 \text{ amperes}$$

Now, find the resistance of the heating element of an electric toaster that carries a current of 5.0 amp on 1 110-volt line:

$$R = V/I$$
$$R = {}^{110}\!/_{5} = 22 \text{ ohms}$$

Exploration 13.1

At extremely low temperatures (near absolute zero), certain materials have zero resistance: That is, they conduct electric current perfectly. These materials are called superconductors. If a superconductor is made into a ring, and a voltage source is applied, a current will flow through the ring. If the voltage source is removed, the current will continue to flow. The temperature at which a substance becomes superconductive is called its *transition temperature.* This temperature ranges from 125 kelvin to within a few thousandths of a degree above absolute zero. Current research into superconductors may one day yield supersensitive electronic devices, very powerful miniature motors, and other major advances in many technological fields.

As Ohm noticed, the resistance of a given conductor usually increases slightly as its temperature is raised. Once the rate of increase has been measured for a given kind of wire, the process may be turned around and temperatures determined by noting the change in resistance of a coil of such wire inserted in the material whose temperature is to be found. Such an instrument is called a resistance thermometer.

Potential Difference in a Circuit

Ohm's law can be applied to a whole circuit or to any part of a circuit, and this often makes it possible to reduce effectively certain groups of resistors to a single unit whose resistance can be computed from those of the individual parts. For instance, suppose that a number of resistors (they may be coils, lamps, heating elements, or any conducting units) are connected in such way that the entire current flows through one after the other. This is called a **series circuit** (Figure 13.5). The potential current will drop along each of the resistors, and the total fall of potential in the whole wire circuit will ɔe the sum of these separate potential differences.

Figure 13.5

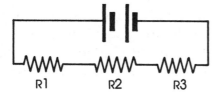

R1   R2   R3

The positive terminal of the battery can be thought of as the highest potential peak in the whole circuit; from here the potential drops as the moving charge goes through one resistor after another, and finally it gets down to the negative battery terminal—which has the lowest potential in the circuit. Inside the battery, chemical

action "boosts" the moving charge back up to the high level, and it goes around the circuit again and again. The close similarity with balls rolling down a slope is suggested by Figure 13.6.

Figure 13.6

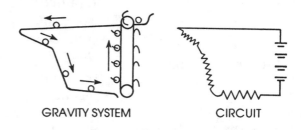

GRAVITY SYSTEM   CIRCUIT

Suppose we have a series circuit like the above, except that it may contain any number of resistors so connected. If the resistance of the first one is called R_1, that of the second one R_2, and so on, then the combined resistance of the whole set (call it simply R) will be merely the sum of the separate ones, or

$$R = R_1 + R_2 + R_3 \ldots \text{etc.}$$

Exploration 13.2

Suppose you want to connect two coils of resistance 2 ohms and 6 ohms into a simple series circuit with a 12-volt battery. What current will the battery deliver, and what is the potential difference across each coil?

The combined resistance of both coils is $2 + 6 = 8$ ohms. Applying Ohm's law to the whole circuit, the current is $I = V/R = {}^{12}/_8 = 1.5$ amp. Ohm's law in the form $V = IR$ may now be applied to the

2-ohm coil alone, giving $V_2 = 1.5 \times 2 = 3.0$ volts. In the same way, for the 6-ohm coil, $V_6 = 1.5 \times 6 = 9.0$ volts. The sum of these two PDs is 12 volts, the voltage of the battery, as it must be.

Suppose you have five 110-volt lamps strung along a circuit that receives line voltage of 550 V. If the lamps are strung in series, the PD across each lamp will be 110 V ($550 = 110 + 110 + 110 + 110 + 110$), the appropriate operating voltage for the lamp. A disadvantage of the arrangement is that if one lamp burns out, the whole set goes out.

To get around difficulties like the one just mentioned, the appliances in a household circuit are connected to make them independent of each other as in Figure 13.7. This is called a **parallel circuit.** The main current, instead of going through one after the other, divides. A part of it goes through each. The separate currents then rejoin and complete the circuit. In this type of circuit, each resistor has the same voltage applied to it—that of the battery. If any unit is disconnected, the remaining ones continue to function as before.

Figure 13.7

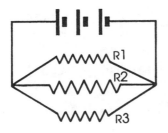

It is found that the combined resistance of a number of resistors connected in parallel is given by

$$\frac{1}{R} = \frac{1}{R_1} + \frac{1}{R_2} + \frac{1}{R_3} \ldots \text{etc.}$$

Here R_1, R_2, etc. stand for the values of the separate resistances, and R is the equivalent total resistance of the set. Be careful: Avoid the mistake of assuming that both sides of the equation can be inverted term by term. All the fractions on the right side must first be brought over a common denominator.

For example, three resistors of 4, 6, and 12 ohms, respectively, are connected in parallel, and a 6-volt battery is applied to the combination. What current is delivered by the battery, and what current flows in each branch? The first thing to do is compute the equivalent resistance of the set. Using the above relation,

$$\frac{1}{R} = \frac{1}{4} + \frac{1}{6} + \frac{1}{12} = \frac{6}{12} = \frac{1}{2}$$

inverting,

$R = 2$ ohms.

Notice that the value of R is less than any of the individual resistance values. This is reasonable, since every conductor added in parallel provides an additional path for the current. The current in the entire circuit is given by Ohm's law as $I = V/R = {}^6/_2 = 3.0$ amp. The current in the 4-ohm coil is $I_4 = V/R_4 = {}^6/_4 = 1.5$ amp. In the same way, the current in the 6-ohm branch is ${}^6/_6 = 1$ amp, and that in the 12-ohm branch is ${}^6/_{12} = 0.5$ amp. The sum is 3.0 amp, as it must be.

Even though series and parallel connections are two very important arrangements of resistors, there are other more complicated hookups that are also used in practice. These often can be handled by using an extension of Ohm's law, but the details may get somewhat complex.

Exploration 13.3

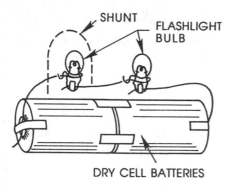

SHUNT   FLASHLIGHT BULB

DRY CELL BATTERIES

Check out the principles you have just learned by using the two cells and bulb of a 3-volt flashlight, together with an extra bulb. First connect the cells and lamps in series, using short lengths of speaker wire whose ends are scraped clean. To make contact at the center terminals of the bulbs or cells, tape the wires on. Notice that the bulbs glow only dimly, since the PD across each is just half what it should be. Now short-circuit one bulb by shunting a piece of wire across it. Most of the current will then go through the "short," which has less resistance than the bulb filament, and so the other bulb brightens up. Put aside the shunt, then connect both bulbs in parallel and observe that they light normally. Finally, put both bulbs in series with a single cell and notice that they are very dim, since each has only one fourth the normal PD applied to it.

Measuring Electric Power

The *power* expended in any appliance—the rate at which it uses electrical energy—is given directly in watts by multiplying the current in amperes by the PD in volts. In symbols,

$$P_{watts} = I_{amp} \times V_{volts}$$

For example, an electric iron that draws 9.0 amp when connected to a 120-volt line would have a power rating of $9.0 \times 120 = 1080$ watts.

Since power is defined as energy divided by time, energy may be expressed as power multiplied by time. This is the way electrical energy is sold, the unit being the **kilowatt-hour.** Thus, when you write a check for your monthly electric bill you are paying for the total amount of electrical *energy* that the company delivered to you during that period. The total energy consumed by a number of appliances is found by adding up the products of power rating and time of use for all of them.

For instance, you use your 1080-watt iron for a total of 20 hours each month, your washing machine (1200 watts) for 12 hours each month, two 60-watt lighting fixtures in your laundry room for 25 hours. If electrical energy costs 14 cents per kilowatt-hour in your town, what is the monthly cost of the appliances used? Remembering that 1 kw = 1000 watts, the total energy used will be $(1.08 \times 20 + 1.2 \times 12 + .12 \times 25) = 39.0$ kwh. At the 14-cent rate, this will cost about $5.40 in all.

Putting Electric Power to Use

The work of moving charges around a circuit may be converted into various forms. Part of the energy may be changed

---

The power of an electrical appliance is the rate at which it uses electrical energy.

---

to mechanical work if there are motors in the circuit; part may be changed into radiation if there are lamps, and so on. Some of the energy will appear in the form of heat. In a wire, for example, the electrons that are made to move through it continually bump into the atoms of the material, delivering some of their energy to them in the form of random heat motion.

Sometimes the heat produced in a conductor is an unavoidable loss, for example, motors or storage batteries. At other times, as in furnaces, heating coils, or cooking stoves, the production of heat is the main purpose of the unit. Then it becomes important to know how to calculate how much heat will be obtained. To do this, recall that the power W (in watts) expended in maintaining a current I (in amp) in a conductor where the PD is V (in volts) is given by $P = I \times V$. If the current flows for a time t (in sec), then the total work done, or energy delivered, W (in joules) is given by $W = I \times V \times t$. But 1 joule is equivalent to $1/4.18$, or 0.24 calorie, and so if all the work done by the source of potential difference is changed to heat in a conductor, the amount of heat produced will be, in calories,

$$Q = 0.24\, I \times V \times t$$

EVERYDAY PHYSICS: 13

"I got so mad I blew every fuse I had!" That's what fuses are for—to prevent an excess of heat from leading to a fire or explosion. If the wires of an electric circuit carry too heavy a current—for example, if you try to run a toaster, a microwave oven, and a hair dryer on the same circuit at the same time—they may become hot enough to burn away their insulation and start a fire.

Fuses and circuit breakers prevent dangerous overloading by providing an intentional "weak spot" in a circuit. Not only household circuits but individual appliances such as amplifiers have such protective devices. A fuse is a strip of wire that has a high resistance and is made of some metal that melts at a relatively low temperature. It is interposed at some point in the circuit. If the current gets too high, the wire in the fuse will melt, breaking the circuit before damage can be done to the appliance or a fire is started.

A circuit breaker is basically a switch with a gap in it. A sudden rise in current causes the electricity to "arc" across the gap and actuate the switch, breaking the circuit. A "blown" fuse must be replaced; a "tripped" circuit breaker can be reset once the overload has been located and corrected.

For example, suppose you want to know how much heat is produced in one minute by an electric iron that draws 4.0 amp when connected to a 115-volt line. If you substitute in the above relation, you get

$$Q = 0.24 \times 4.0 \times 115 \times 60 = 6{,}624 \text{ cal.}$$

For some purposes it is more convenient to have the quantity of heat given in terms of current and resistance, rather than current and voltage. Using Ohm's law, we can substitute $I \times R$ for V in the formula, getting

$$Q = 0.24 \, I^2 \times R \times t$$

This expression shows, for instance, that if a number of resistors are connected in series (same current in each), the greatest amount of heat will be produced in the one having the highest resistance. Since the resistance of a conductor changes with temperature, the heat produced by the current will change the value of R, and care must be taken to use the value that corresponds to the temperature reached in any particular case.

The electric arc, which may be used for lighting, for heating certain types of industrial furnaces, or for welding metals together, utilizes the heat evolved by the current. So do a variety of household appliances such as waffle irons, heating pads, coffee makers, electric blankets, and many more.

Testing Your Knowledge

13.1 How strong is the average current, in ampheres, in a lightning flash lasting 0.0002 sec if 1 coulomb of charge passes?

13.2 Could you make a voltaic cell by placing two strips of zinc in an acid solution?

13.3 About how many dry cells would have to be joined together in order to have the same total PD as a storage battery consisting of 9 cells?

13.4 When a storage battery is in use, sulfuric acid is being removed from solution. Sulfuric acid has a greater specific gravity than pure water. Where would you expect a hydrometer to float higher—in the liquid from a discharged cell or from a fully-charged one?

13.5 What is actually "stored" in a storage battery—electricity, kinetic energy, chemical energy, or heat? Explain.

13.6 If both the diameter and the length of a copper wire are doubled, what effect does this have on its resistance?

13.7 Christmas tree lights are usually connected in a series of 8 lamps when used on a 120-volt line. If the current in each is 0.2 amp, what is the resistance of each lamp?

13.8 An appliance to be used on a 120-volt line has a resistance of 25 ohms. If the current is to be kept down to a value of 2.0 amp, how big a resistor must be connected in series with the appliance?

13.9 Two resistors connected in parallel are joined to a battery. If one of the resistors has 3 times the resistance of the other, compare the currents in the two. What fraction of the *total* current goes through each?

13.10 A 3-ohm coil and a 6-ohm coil are connected in parallel and the combination is joined in series with a 2-ohm coil and a 12-volt battery. Find the current in the 2-ohm coil.

13.11 In the last problem, find the current in each of the other coils.

13.12 A 30-watt automobile lamp is supplied by the 12-volt storage battery. What is the resistance of the hot filament of the lamp?

Magnetism

Magnetism is the ability to attract iron and certain other metals that have a molecular structure similar to iron (**ferrous metal**). The ancient Greeks knew that a specific kind of rock could attract iron. In ancient China, splinters of magnetic rock were used to guide ships, because it was found that, if hung by a thread, they would always set themselves in a north-south direction. Much more recently, the nature of magnetism and the existence of magnetic poles was explained and put to use, opening the door to the "electric world" we know today.

How Magnets Work

Magnetite is a magnetized iron oxide commonly known as lodestone. An ordinary bar of iron or steel can be made magnetic by stroking it—always in the same direction—with either end of a piece of magnetite. Once this is done, the metal bar will behave just like magnetite. Besides iron and steel, only a very few other materials can be noticeably magnetized: the chemical elements nickel and cobalt, and certain special alloys (manmade combinations of metals).

One end of a magnet will always swing toward the north if freely suspended; the other end will always swing south. The two ends are called the north (N) and south (S) poles of the magnet respectively.

The directional property of magnets that people have used in compasses from ancient times until today comes from the polar quality of magnetism. When a magnetized bar is dipped into a heap of iron filings or chips, heavy tufts cling to the bar near each end (Figure 14.1). These ends are called **magnetic poles.** The pole that turns toward the north when the bar is suspended is called a north pole (or simply N); the end that swings toward the south is called the south (or S) pole.

Figure 14.1

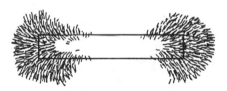

An ordinary piece of iron is always attracted by either pole of a magnet, but by experimenting with a magnet and a compass needle you will find that *like poles repel each other and opposite poles attract.* By experimenting with different magnets at a number of different distances apart, scientists established that the force of attraction or repulsion between poles is directly proportional to the strengths of the poles and inversely proportional to the square of the distance between them. Any action of one magnet against another can therefore be described as the net effect of the forces between the pairs of poles.

If a bar magnet is cut in two, a new pole will exist near each cut end, opposite in kind to the pole at the other end of the piece. No matter how many times the cutting process is carried out, this is found to happen; *it is impossible to have a piece with only one pole on it.* This suggests that when the limit is reached in such a cutting-up process, each piece would still be equivalent to a bar magnet having two poles. In fact, many observations can be explained by the assumption that magnetic materials are composed of tiny individual magnetic units.

In an unmagnetized piece of iron, for example, these units can be thought of as being arranged in random directions. When the bar is magnetized—say, by stroking with a magnet—these small units become lined up, all in the same general direction. The poles of the whole bar represent merely the outside effect of all the individual magnetic units acting together. These tiny units may consist of individual atoms or molecules, or of groups of atoms

Like magnetic poles repel; unlike poles attract. The force of the attraction is directly proportional to the strengths of the poles and inversely proportional to the square of the distance between them.

Magnetism conveyed to another metal object by a magnet is called induced magnetism, and the process is called magnetic induction. Induced magnetism is temporary.

lined up to form small crystals called magnetic domains.

If a magnet is heated red hot and allowed to cool again, it will no longer be magnetic—the jostling of the molecules will have knocked the magnetic units out of their former alignment.

Figure 14.2

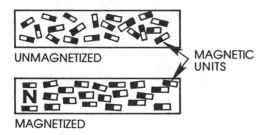

The theory also explains what is called **induced magnetism.** A steel magnet will pick up several tacks or small nails, chain fashion. If the uppermost nail is carefully held in a clamp and the magnet then removed, the whole chain goes to pieces. Each nail attracted the ones next to it only so long as the magnet was near, and we say that magnetism was induced in the nails by the permanent magnet. It kept the units of the iron lined up. Soft iron becomes only temporarily magnetic under the influence of the nearby permanent steel magnet, but a piece of hard steel retains much of its magnetism afterward.

Magnetic Fields

A permanent magnet will exert force from some distance away. The space around a magnet in which its effects are felt is called its **magnetic field.** Although a magnetic field is strongest near the magnet or magnets that cause it, it extends out indefinitely into space. Within the magnetic field, force is exerted in curved patterns known as magnetic lines of force (Figure 14.3).

Figure 14.3

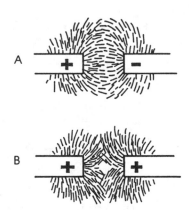

Early in the last century, the great English scientist Michael Faraday saw that the lines of force, while purely imaginary (like light rays, or the earth's equator), could help his thinking about the behavior of magnets if he assumed that they acted like stiffened rubber bands. For instance, the filings show that the lines of force be-

---

The space around magnets in which their force is felt is called a magnetic field; the force is exerted in patterns called magnetic lines of force.

---

tween two poles of opposite kind go across from one pole to the other. If they were really stretched rubber bands, this would tend to pull the poles together, which is what actually happens. For two poles of the same kind, the lines bend away from each other, and if they are assumed to act like stiff fibers, this would account for the fact that the poles tend to push apart. As to direction, the lines are assumed to go out from N poles and come in to S poles.

Exploration 14.1

You can buy inexpensive toy horseshoe magnets or small but powerful bar magnets in a hardware store or electronics hobby shop. Lay a card on top of one or more of these magnets, scatter some iron filings uniformly over the whole card, then tap it with a pencil. The filings become magnets by induction and arrange themselves in curved lines much like those in Figure 14.3. These curves are called **magnetic lines of force.**

The Earth's Magnetic Poles

Why does a compass needle line up in a north-south direction? It must be that the earth itself is surrounded by a magnetic field. What causes the whole earth to act as a magnet is not completely understood. Part of the effect seems to be

due to strong electric currents in the earth's core, and the rotation of the globe may play some part. At any rate, the character of the field is about what would be expected if there were a huge bar magnet inside the earth. Since the pole of a compass that points northward was designated an N pole, it follows that the imagined earth magnet has its S pole in the southern part of the globe (Figure 14.4).

Figure 14.4

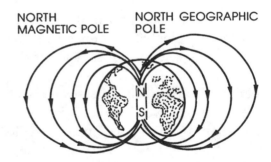

NORTH MAGNETIC POLE

NORTH GEOGRAPHIC POLE

At any place on the surface of the earth in, say, the northern hemisphere, the lines of the field dip down into the ground at an angle. The place where the lines go straight down is called the *north magnetic pole*, and is located in northern Canada nearly 1500 miles from the geographic north pole. There is a corresponding *south magnetic pole* in nearly the opposite location. Because the magnetic poles and the geographic poles are not at the same place, compass-indicated north differs considerably in most places from true north. In the northeastern part of the United States, the compass points almost north-northwest, while on the Pacific Coast it points north-northeast. Navigators must have charts showing how much to correct this error of the compass, but the field is found to change slowly over a period of years, so that the government must issue new maps from time to time. At irregular intervals there also occur sudden, violent changes in the field, which may seriously affect telegraph, radio, and TV communication. These "magnetic storms" are believed to be connected with increased activity of the sun.

Electrical Fields

Some of the most important technical applications of electricity depend on the fact that a current produces a magnetic field in its neighborhood. This connection between magnetism and electricity was discovered by the Danish physicist H. C. Oersted. He noticed that a compass needle placed just below a wire carrying a current would take up a position nearly perpendicular to the wire while the current was flowing. When the direction of the current was reversed, the needle again set itself at right angles to the wire, but with its ends reversed. The effect lasts only while the current flows. It is not due to the wire as such (copper is nonmagnetic),

but in some way to the existence of the current itself. In fact, currents in solutions or in gases and charges streaming across a vacuum are found to give the same effect.

The lines of magnetic force that mark out the field due to a current in a straight piece of wire are found to be circles that go around the wire in one direction (Figure 14.5). This can be checked by carrying a small compass needle around the wire or by scattering iron filings on a card through which the wire passes. The field is strongest near the wire and gets weaker as you go farther out in any direction. If the current is reversed, the lines are again circles, but go around in the opposite sense.

Figure 14.5

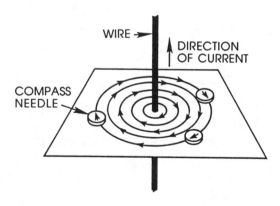

The direction of the field in relation to the direction of the current carrying it can be simply designated as + or −. To do so, we must first define what is meant by the direction of current flow in a wire. Long

---

Electric current produces a magnetic field in a circular pattern around a straight current such as that in an electrical wire.

before the discovery of electrons, the current in a wire had always been taken to flow from the + terminal of a battery to the − terminal. You may still see it used this way in some books. However, we now know that electrons are negatively charged, so they must flow from negative to positive. In this book we will say that the direction of a current is the direction of the electron flow, that is, from − to +.

The simple rule for the relationship between the direction of current in a wire and the magnetic field is known as the **left-hand rule.** Imagine that you are grasping the wire with your left hand, your thumb extended in the direction of the current. Your fingers will then be encircling the wire in the direction of the magnetic lines of force.

Figure 14.6

Electromagnets

The French physicist Ampère found that the magnetic effect of current in a wire could be greatly increased by winding the wire in the form of a spinal coil, or **solenoid.** The effects of the many turns add up to give a field exactly like that of a bar magnet, and the lines can be followed even inside the coil (Figure 14.7).

Figure 14.7

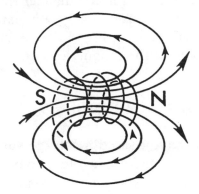

The magnetic strength of such a coil may be increased hundreds and even thousands of times by placing a soft-iron core inside it. The device is then called an **electromagnet** (Figure 14.8). It has the advantage over a permanent magnet that it can be made much stronger and that its strength can be controlled and its polarity reversed by suitably changing the current in the coil.

Figure 14.8

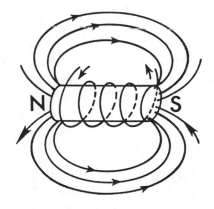

There is a definite relationship between the direction of the current in the coil and the direction of the magnetic field. You can use the left-hand rule to find the direction of the current in the coil. When you grasp the coil in the manner that was shown in Figure 14.6, your extended thumb will be pointing in the direction of the N pole of the magnetic field.

Uses of Electromagnets

If you have passed a salvage yard and seen a crane picking up old auto bodies or piles of scrap, you have seen an electromagnet in action. Lifting-magnets strong enough to hold many tons are used to load steel rails or bars, machine parts, and scrap iron. The load is picked up or released by opening or closing the switch that controls the current in the coils. Some electromagnets can lift as much as 200 pounds for each square inch of pole face.

A **relay** is an electromagnetic device that allows a weak current to open and close a circuit in which a heavier current flows. Basically, it is a switch that can be operated from a distance. A sensitive relay may operate on current as small as a millionth of an ampere.

Electromagnets are used in everyday appliances like telephone receivers and loudspeakers, and in complex research devices such as mass spectographs and cyclotrons.

Forces Operating on an Electric Current

Experience shows that a current-carrying wire placed in a magnetic field is acted on by a sidewise force. As an example, suppose that, in Figure 14.9a, a wire extends in a direction perpendicular to the lines of the field of the magnet. Then, with the field and current as shown, the wire is found to be pushed to the right. The three directions—current, field, and force—are mutually perpendicular, like the three edges of a brick that go out from a corner. This mechanical effect can be thought of as the action of the field of the magnet on the field produced by the current in the wire, and the idea of lines of force will give its direction.

In Figure 14.9b, both the lines of the magnet's field and those of the current in the wire have been drawn. But two sets of lines can always be combined into a single set, for at any point the two forces themselves can be combined into a single resultant. At any point to the right of the wire, the two fields are in opposite directions, and so partially cancel; to the left of the wire, the two are in the same direction and reinforce each other. The combined field is shown in Figure 14.9c. Remembering that the lines tend to act like stretched bands, the effect of a field of this shape would be to force the wire over to the right, as shown.

Figure 14.9

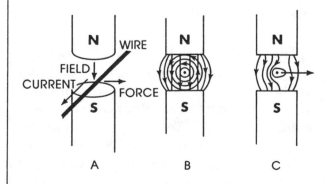

A   B   C

Metering Electricity

The commonest forms of **ammeters** and **voltmeters** operate on the basis of the forces acting on current-carrying wires in a magnetic field. The field is usually that of a strong permanent magnet, and the current to be measured is passed through a rectangular coil mounted on good bearings (Figure 14.10). A stationary soft-iron core inside the coil acts to concentrate the field. When current flows in the coil, the action of each wire that extends in the direction perpendicular to the page is like that described above, and the net effect is to turn the coil in one direction on its axis. This turning is opposed by a pair of hair springs, and since the magnetic forces are proportional to the current, the amount that the coil turns will be a measure of this current.

If such an instrument is to be an ammeter, most of the main current in the circuit must bypass the coil and go through the low-resistance shunt. But a definite fraction will go through the coil, and the scale can be marked to read directly the total current passing through the meter. On the other hand, if the instrument is to be a voltmeter, it must have a high resistance, so that the current it draws is not appreciable. In this case a stationary coil of high resistance is connected in series with the moving coil. The movement of the coil is determined, as above, by the current flowing in it. By Ohm's law, this is proportional to the applied PD, and so the scale can be marked directly in volts. Moving-coil instruments can be made sensitive enough to respond to currents as small as a hundred-billionth of an ampere.

Figure 14.10

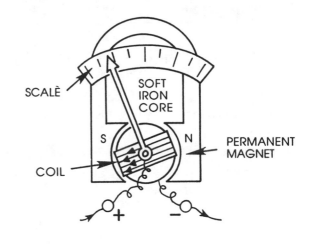

Exploration 14.2

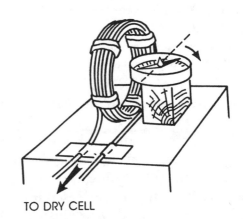

TO DRY CELL

You can make a simple current-indicating instrument having a fixed coil and a moving needle. Wind about 25 turns of bell wire on a small glass or bottle, leaving about a foot of straight wire at each end. Slip the windings off, tape them together, and mount the coil in an upright position on a piece of wood. Place a pocket compass opposite the center of the coil and set the arrangement in a north-south direction,

---

A device that steadily converts electrical energy into mechanical energy is called an electric motor.

---

so that the face of the coil is parallel to the compass needle. When the coil is connected to a dry cell, the flow of current will be indicated by the swinging aside of the needle. Reversing the battery connections makes the needle swing the other way. Check the coil rule for this case.

Motors

If a current-carrying coil is allowed to turn freely in a magnetic field as in the current meters described above, it will acquire kinetic energy. If this turning could be made to continue we would have a steady conversion of electrical into mechanical energy. Any device for doing this is called an electric motor. For simplicity, suppose the coil consists of only a single loop. If continuous turning is to take place, the current can no longer be led into and out of the coil by fixed wires; instead, this is done through a split ring, called a commutator, on which sliding contacts (brushes) bear.

When the current is going through the loop in the direction shown in Figure 14.11a, the loop will turn until its plane is vertical. At that moment, however, the current through the loop is automatically reversed by the switching around of connections as the commutator gaps pass the brushes. This reversal lets the coil make another half turn, when reversal occurs

Figure 14.11

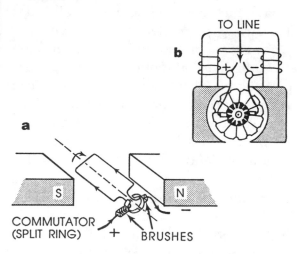

again, etc., with the result that the coil turns continuously in one direction.

A practical motor, such as the "starter" of an automobile engine or the motor of a battery-operated appliance, differs in design from the simple device just described. The single loop is replaced by a set of separate coils wound into recesses spaced around the curved surface of a soft-iron core, and the commutator has two opposite segments for each coil. The field magnet is usually an electromagnet, and all or part of the current supplied to the motor is passed through its windings, (14.11b).

Commercial motors convert about 75 percent of the electrical energy supplied to them into mechanical work. Some are very small and are battery driven, such as the motor in an electric shaver. Others are very large, such as the electric motors that drive monorails or ships.

Testing Your Knowledge

14.1 Of the following, the one that will be attracted by *either* pole of a bar magnet is
   a. the N pole of another bar magnet.
   b. the S pole of another bar magnet.
   c. an unmagnetized piece of nickel.
   d. bits of broken glass.

14.2 Two identical bar magnets are placed side by side, with like poles together. This combination will
   a. pick up bits of steel but not iron.
   b. pick up more iron filings than one such magnet.
   c. fail to attract a cobalt rod.
   d. repel iron nails.

14.3 If one end of a soft-iron rod is held close to the S pole of a strong magnet,
   a. the rod will be a strong magnet after the bar magnet is removed.
   b. there will be, temporarily, an S pole near the other end of the rod.
   c. the rod will spring away as soon as it is let go.
   d. the far end of the rod will not attract iron filings.

14.4 In flying from Chicago to Calcutta via the North Pole, the plane's magnetic compass would point
   a. northward at all times.
   b. southward at all times.
   c. northward part of the time.
   d. directly toward the geographic pole at all times.

14.5 A magnetized needle free to pivot in any direction will come to rest in a horizontal position when at
   a. the north magnetic pole.
   b. the south magnetic pole.
   c. a place about midway between the north and south magnetic poles.
   d. a place about 1500 miles from the north magnetic pole.

14.6 The wires leading to a filament lamp do not become as hot as the filament itself because the filament has a greater
   a. length.
   b. diameter.
   c. resistance.
   d. current flowing in it.

14.7 When a 100-watt water heater is allowed to run for 5.0 min, it raises the temperature of ½ pint (225 gm) of water by
   a. 0.53 C°.   c. 72 C°.
   b. 32 C°.   d. 0.22 C°.

14.8 The century in which Oersted discovered that a magnetic effect can be produced by an electric current was the
   a. eighteenth.   c. twentieth.
   b. nineteenth.   d. sixteenth.

14.9 A vertical wire carries a heavy current flowing from bottom to top. A compass needle placed on a table just east of the wire will point
   a. north.   c. east.
   b. west.   d. south.

14.10 A spiral coil is wound on an upright post, with the current going around it in a clockwise direction as seen from above. Then it is true that the lines of force will
   a. begin and end on various turns of a wire.   c. decrease in number if the current is increased.
   b. concentrate most just beyond the ends of the coil.   d. enter the coil at the top.

14.11 Two current-carrying wires are side by side a short distance apart. What effect do they have on each other? (*Hint:* Either current can be considered to be in the magnetic field due to the other.) Make a sketch showing the direction of the force on each wire (a) when the two currents are in the same direction and (b) when opposite.

14.12 If an ammeter whose coil has a resistance of 0.09 ohm is used with a shunt of resistance 0.01 ohm, what current actually will flow through the coil when 10 amp flows into the instrument?

14.13 If a voltmeter has a coil of resistance 0.1 ohm and a series resistor of 500 ohms, what will be the PD across the coil itself when the instrument is connected to the terminals of a 10-volt battery?

14.14 What is the effect, if any, on the direction of rotation of the motor diagrammed in Figure 14.11 of reversing the connections to the line?

Electromagnetic Induction

Once it was known that electricity can produce magnetism, scientists began looking for ways to produce electric currents by means of magnetism. Almost simultaneously, Michael Faraday in England and Joseph Henry in the United States conducted the experiments in which magnets and coils were used to produce electric current. Their discoveries made the commercial development of electricity possible.

How Electromagnetic Induction Works

In one experiment, Faraday connected a coil directly to a meter and found that, when one pole of a bar magnet was moved quickly toward the coil, the meter registered a momentary current (Figure 15.1). When the magnet was jerked away, another brief current was registered, but in

Producing currents through the use of coil and magnets is called *electromagnetic induction.*

the opposite direction. Faraday found that the magnitude (strength) of the current increased with the strength of the magnet, its speed of motion, and the number of turns of wire in the coil.

Figure 15.1

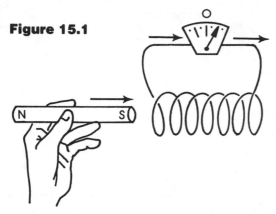

Another experiment showed that a meter connected to a coil showed current the instant that current was started or stopped in an entirely separate circuit nearby. Thus, when the primary circuit containing coil one is closed, the nearby secondary circuit containing coil two produces a momentary "kick" of current. So long as a steady current flows through the primary circuit, nothing further happens, but if the primary circuit is interrupted, there is a momentary impulse of current in the secondary, opposite in direction to the original current.

Faraday described what was happening in terms of **magnetic flux,** the total number of lines of force that pass through any closed loop located in a magnetic field.

When a current is induced in a coil, a change of the flux occurs. For example, in the experiment shown in Figure 15.1, the lines of force within the coil move with the magnet, changing the flux through the various turns of the coil.

In Figure 15.2, closing the switch makes coil CP a magnet, and lines of force spring up all around it. Some of these lines thread through the turns of CS, where no flux existed before. So long as the current in CP remains constant, no change of flux will occur in CS, and so no induced current will occur. However, opening the keys (interrupting the circuit) makes the flux in CP disappear, meaning that the flux through CS changes, and current is again induced.

Figure 15.2

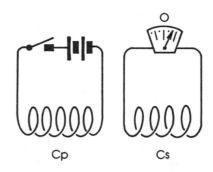

Cp   Cs

Some experiments are done more easily with a single wire rather than a coil. In Figure 15.3, for example, a wire has been

Magnetic flux is the name for the total number of lines of force passing through any closed loop (circuit) located in a magnetic field.

connected into a complete circuit containing a meter. By moving the meter horizontally near one pole of a stationary magnet, a potential difference can be shown to exist between the ends of the wire. The current in this instance is said to be caused by the cutting lines of force by the wire. Even so, the motion of the wire changes the flux through the circuit, and this change in flux is the cause of the induced current.

Figure 15.3

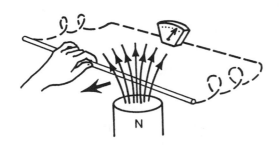

Lenz's Law

Faraday's experiments showed that there is a direct relation between the direction of an induced current and the direction of the action that causes it. If the magnet is pushed toward the end of the coil, the flux through the turns increases and a current is induced in the coil. This current magnetizes the coil.

Suppose the direction of the induced current is such that the upper end of the coil becomes an S pole. In that case, the attraction between the S pole of the coil and the N pole of the bar will help the movement of the bar magnet along. The bar magnet need no longer be pushed into the coil; its motion would reinforce itself, producing unlimited amounts of both mechanical and electrical energy without the expenditure of further effort.

The law of conservation of energy tells us this will not happen. Instead, the top end of the coil becomes an N pole like the bar. Its repulsion now *opposes* the movement of the bar magnet. Similarly, when the magnet is pulled away from the coil, the induced current must be in the opposite direction, making the top end of the coil an S pole whose attraction opposes the removal of the bar.

Experience bears these facts out. The direction of an induced current is always such that its magnetic field opposes the operation that causes it. This generalization is called **Lenz's law** in honor of its discoverer, Heinrich Friedrich Lenz.

Generators

Before electromagnetic induction could be made to produce more than temporary, weak current, continuously operating **generators** had to be designed. The essential parts of a generator are the same as those of an electric motor: a coil (electromagnet), a magnetic field in which the coil can be rotated, and some means for connecting the coil to an outside circuit. In fact,

---

Lenz's law: the direction of an induced current is always such that its magnetic field opposes the operation that causes it.

with slight adjustments, the same machine may be used as either a motor or a generator. If a current from some outside source is passed into the coil, it rotates and acts like a motor; that is, it converts electrical force into mechanical force. If the coil is mechanically turned, as by an engine or a water-driven turbine, an induced current results; that is, the machine converts mechanical energy into electrical energy.

Alternating Current (AC)

Figure 15.5 shows several different positions a rotating coil would take while inducing voltage. (The coil is not moving sideways; the positions a to d represent the same coil at different times.) The voltage graph at the bottom of the figure shows the directions of the current. As the coil is rotated, the nature of the induced voltage changes as Lenz's law would predict:

○ At a, the plane of the coil is perpendicular to the magnetic field.
○ Turning the coil onward in the direction shown makes its right-hand face an N pole and the other an S pole, so that the forces between the poles of the coil and the poles of the field magnet act to *hinder* the motions.
○ The direction of the induced voltage (or the current it gives rise to) will be as indicated by the arrows.

○ At b, the coil has turned through a right angle, and the voltage has increased to its maximum value because its horizontal wires have been cutting more and more squarely across the lines of force.
○ As the coil approaches position c, the voltage falls back to low values, and when the coil passes this point the voltage actually reverses.
○ During the next quarter turn (d), the voltage gets larger and larger in this reverse direction.
○ In the final quarter turn, it lapses back to zero and the whole cycle of events starts again (e).

Figure 15.4

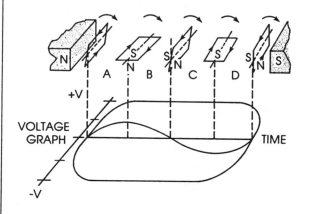

An alternating current (AC) is the kind that naturally results from the continued turning of a coil in a fixed magnetic field.

The voltage graph at the base of the diagram in Figure 15.4 shows that the voltage is alternating; that is, it goes first in one direction and then in the opposite. If the coil is now connected to an outside circuit by means of slip rings and brushes, the current furnished to this circuit will be **alternating current (AC)**. The number of complete cycles equals the number rotations per second of the coil. Thus an alternating current is the kind that naturally results from the turning of a coil (electromagnet) in a fixed magnetic field.

In all but the smallest generators, the field magnets are electromagnets rather than permanent ones. Figure 15.5 represents a typical AC generator.

Figure 15.5

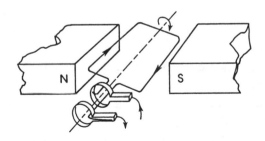

Direct Current (DC)

Alternating currents are well suited to many purposes such as heating and lighting. Other uses, such as electroplating or the charging of storage batteries, require **direct current (DC),** which always flows in one direction. To obtain direct current from a generator of the kind just described, the slip rings must be replaced by a commutator. The switching-over action of this device has the effect of reversing alternate loops in the output wave (Figure 15.6). The current in the outside circuit is now always in one direction. However, it is still far from steady, rising to a maximum and falling back to zero every half-cycle.

Figure 15.6

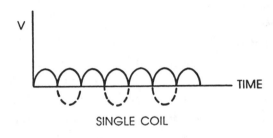

SINGLE COIL

A more constant current can be achieved if the rotating part of the generator is wound with several coils set at different angles to each other (Figure 15.7). The output of each coil then reaches its maximum while others are at intermediate positions, and the combined output shows relatively little variation. The figure shows the effect of adding the outputs of three equally spaced coils. With more coils, hardly any "ripple" would remain.

Direct current (DC) always flows in one direction; for certain purposes such as charging storage batteries, DC is preferable.

Figure 15.7

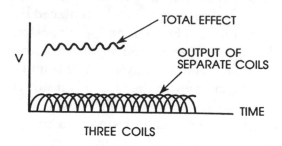

Back Voltage

When an electric motor is in operation, its rotating windings cut the lines of force of the field magnet, and so the motor will at the same time act as a generator. The direction of the induced voltage will be opposite to the PD that supplies current to the motor, and so is called **back voltage.** The back voltage increases with the speed of rotation, and the difference between the applied PD and the back voltage at any time determines how much current actually enters the motor. When a motor is just starting up, its back voltage will be very small because the rotation is slow. Without this back voltage, a large current would surge through the windings and perhaps burn them out. For this reason a starting box, consisting of a chain of several resistors, is placed in series with the motor. As the motor picks up speed and its back voltage comes up to the operating value, these protective resistors are cut out of the circuit one after another, until finally the full line voltage is applied to the motor.

Transformers

One reason that alternating current is widely used is that voltage and current values may be readily and efficiently changed by the use of a device called the **transformer.** In principle, the pair of coils in Figure 15.8 is a transformer. Any change in the current in the primary coil induces a voltage in the secondary. If an alternating current is supplied to the primary, there will be a corresponding variation of magnetic flux through the secondary. As a result, an alternating current of the same frequency will be induced in the secondary. In the United States, the frequency used on domestic power lines is 60 cycles—that is, the current makes 60 complete vibrations per second.

Simple "air-cored" transformers are used in radio and TV circuits, but for power transmission, the two coils are wound on a closed ring of special steel that increases and concentrates the magnetic flux. With this design, the flux at any time is the same for all turns, and the result is that the voltages in the two coils are proportional to the number of turns, or

$$\frac{V_s}{V_p} = \frac{n_s}{n_p}$$

where V_s = voltage in the secondary, V_p = voltage in the primary, n_s = number of turns in secondary, and n_p = number of turns in primary. If there are more turns in the secondary than in the primary, the voltage of the secondary will be greater than the primary voltage and the device is · called a "step-up" transformer; if the other way around, a "step-down" transformer.

Figure 15.8

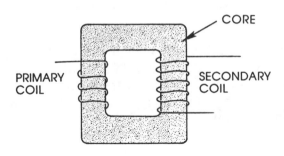

PRIMARY COIL

SECONDARY COIL

CORE

When electric power is to be used at great distance from the generator, it is transmitted in the form of high voltage AC, for the following reason: the heat loss in a line is proportional to I^2R, so if the losses are to be reduced the current should be as small as possible. With a given power, this means making the voltage high, since $P = IV$. These current and voltage changes can be made economically only through the use of AC, which permits the use of transformers. In a power plant the generator voltage may be about 10,000. A transformer steps it up to perhaps 230,000 and puts it on the transmission line. At the edge of a city, a step-down transformer may reduce the PD to about 2300 and small step-down transformers located on power-line poles throughout the city then reduce it to a safe value of about 110 volts for use in homes.

There are no moving parts in a transformer, and when properly designed the energy losses may be as low as 2 percent. This means that, practically, the same amount of power is developed in each coil. As in the case of direct currents, the power developed in either one is given by current multiplied by voltage, so that $I_p V_p = I_s V_s$, or $I_s/I_p = V_p/V_s$. Combining with the relation above, we have

$$\frac{I_s}{I_p} = \frac{n_p}{n_s},$$

so that the currents in the two coils are inversely proportional to the number of turns in each.

For example, the primary and secondary coils of a power-line transformer have 50 and 25,000 turns respectively. Neglecting losses, if AC of effective voltage 110 is supplied to the primary, what will be the voltage in the secondary? The relation in the previous formula gives $V_s = V_p\, n_s/n_p$ $= 110 \times 25,000/50 = 55,000$ volts.

Exploration 15.1

Many electrical devices transform one kind of energy into another. In the telephone, for example, sound vibrations (compression waves) are converted into electrical current for transmission and then reconverted to compression energy at the receiving end. In tape recording, a magnetic coil transfers sound signals to a plastic tape coated with magnetic oxide for permanent storage in a pattern representing the original sound waves. In digital recordings, compression waves are translated into coded (digitized) electrical signals, which are read and reconverted into sound when the recordings are played back.

Testing Your Knowledge

15.1 A closed wire hoop is slid around in various directions while lying flat on a table located in a uniform vertical magnetic field. Is any current induced in the hoop? Explain.

15.2 Feeble induced voltages, due to motion in the earth's magnetic field, are to be expected in the axles of a moving railroad car. If a train is moving northward, in what direction will the induced voltage be?

15.3 What is the effect on the voltage delivered if the speed of rotation of a generator is increased?

15.4 Find the ratio of the number of turns in the primary to the number in the secondary of a toy transformer that steps down the 110-volt house current to 22 volts for operating a model electric railroad.

15.5 In a spot welder, where very large currents are needed for producing the required heat, a transformer having a 100-turn primary and a 2-turn secondary is used. If a current of 1 amp is admitted to the primary, what current can be obtained from the secondary?

15.6 A generator turns very easily when no current is being drawn from it, but becomes very hard to turn as soon as the switch connecting it to the outside circuit is closed. Explain.

15.7 The rotating coils of a motor that is used on a 50-volt line have a total resistance of 2 ohms. What current flows in them when the back voltage of the motor amounts to 45 volts? If connected directly to the line while standing still, how large a current would flow in the windings? What might then happen?

CHAPTER 16

Electromagnetic Radiation

KEY TERMS FOR THIS CHAPTER

cathode ray
electron
photoelectric effect
vacuum tube

semiconductor
transistor
integrated circuit
amplitude modulation (AM)

frequency modulation (FM)
cathode ray tube (CRT)
electron microscope

Light waves, which are one form of electromagnetic radiation, were discussed in Chapters 10 and 11. Other types of electron beams (electromagnetic radiation) have many uses. Both pure and applied science, and especially communications, have been drastically affected by electron-beam technologies based on discoveries about atomic particles that were made nearly a century ago.

Electron Beams

Early researchers tried passing a high-voltage direct current through a tube containing air at very low pressure. Ordinarily, gas glows when a current is passed through it; this is the principle, for example, of neon lighting. However, it was found that, when the pressure of the gas is reduced to about $1/100,000$ of normal at-

---

Cathode rays, which cause fluorescence in glass in the presence of a vacuum, consist of negatively charged electrons.

---

mospheric pressure, its glow disappears. Instead, the glass of the tube begins to glow with a greenish light.

Something that appeared to be coming out of the negative end of the tube in straight lines caused the glow, and early experimenters called these emanations **cathode rays.** Besides causing fluorescence (glowing) of the glass, these rays were found to deliver energy by impact, so it was concluded that they must consist of streams of particles. Also, they could be bent aside by applying a magnetic field, which meant that they must be equivalent to an electric current. This and other evidence led to the conclusion that cathode rays consisted of *negatively charged* particles.

Experiments by J. J. Thomson of England eventually showed that regardless of the gas used in the tube or other variables, each cathode ray particle had a measured mass of $1/1840$ of the mass of a hydrogen atom. From this, Thomson concluded that the so-called cathode rays consisted of subatomic particles that are part of every atom. He named these particles **electrons.** Other workers obtained more precise measurements of the charge and mass of electrons. It was ultimately established that cathode rays consist of electrons, each of which carries a charge of negative electricity.

Electrons inside a tube of gas at a very high vacuum attain great speeds because they can travel almost without hindrance. The German physicist Wilhelm Roentgen found that when a cathode ray beam strikes the end of the tube, it produces a type of beam that Roentgen called an X ray. Later researchers showed X rays to be a kind of electromagnetic wave with a very short wavelength. X rays can penetrate many kinds of material, including human tissue, and are therefore widely used in medical diagnosis and in physics research. The entire medical field of radiology (roentgenographic examination) originated with Roentgen's discoveries.

Photoelectric Effect

We have already discussed how electrons can be freed from a solid substance by means of heat. Another way of freeing electrons in quantity is by shining light on suitable materials, creating what is called the **photoelectric effect.** In a typical photocell, light of a suitable wavelength is allowed to fall through glass onto an inner coating of a metal such as potassium. When light hits the metal, electrons stream out and can be attracted to a positive terminal, giving rise to a current

Figure 16.1

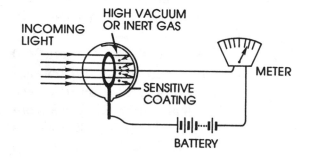

The photoelectric effect occurs when electrons are freed from a solid substance by shining light on suitable materials.

---

in the outside circuit (Figure 16.1). The strength of the current is proportional to the intensity of the incoming light, so that photocells can easily be used to measure light for photography, astronomy, architectural design, or other uses.

Smoke detectors and burglar alarms use photocells; in this case, the device is set off when light fails to reach the photocell. Some types of photocells produce their own voltage, needing no batteries. Solar cells used to power calculators and spacecraft utilize this principle. The sound track on a motion picture allows a succession of light and dark places on a strip at the edge of the film to create variations of current that are converted into sound energy and amplified.

Vacuum Tubes

When a piece of metal or some kinds of metallic oxides are placed in a vacuum and heated, some of the free electrons attain speeds fast enough to enable them to break away and form a cloud near the surface of the container. This "boiling out" of electrons is called thermionic emission. A vacuum tube containing a heated emitter and a plate can be used to rectify an alternating current; that is, to change AC to DC. A vacuum tube of this kind is called a diode.

Figure 16.2 shows what happens inside a vacuum tube seen in cross-section. In this case, a metal filament is the emitter. A current is passed through it to heat it. If a source of alternating voltage is connected between filament and plate, electrons will move from filament to plate whenever the plate is positive with respect to the filament, but not when the potential difference (PD) is the other way around.

In this way, applying an AC voltage to the tube results in an interrupted DC to the outside circuit. By using two diodes, both loops of the AC cycle can be used, and by passing the output through suitable circuits consisting of coils and capacitors, the pulsations can be smoothed out almost completely.

Figure 16.2

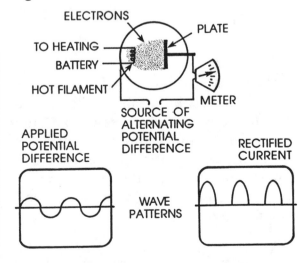

The amount of energy produced by the photoelectric effect increases with the intensity of the light.

A vacuum tube in which a wire mesh, or grid, is inserted between the emitter and plate is called a triode (Figure 16.3). With this arrangement, very small differences in potential can produce large changes in the electron current. By using a series of triodes, several stages of amplification can be achieved.

Figure 16.3

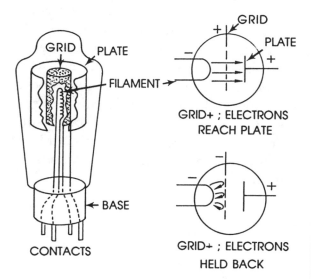

GRID

PLATE

FILAMENT

BASE

CONTACTS

GRID

PLATE

GRID+ ; ELECTRONS
REACH PLATE

GRID+ ; ELECTRONS
HELD BACK

The ability to convert AC to DC and to amplify signals by means of the vacuum tube led to major advances in communication, including radio and television. Vacuum tubes require considerable power, however, and they have been replaced in many applications by semiconductors.

Semiconductors

A **semiconductor** is a solid-state device that can be used to transmit and emplify electronic signals. There are three basic kinds of solid-state materials:

○ Metals, which have a low resistance (current flows easily through them)
○ Semiconductors, which have medium resistivity
○ Insulators, which have high resistivity

Some materials are *p* material (having a positive free charge); others are *n* (having a negative free charge). If a *p* material and an *n* material are placed near each other, electrons will flow from *n* to *p*. That is the functional principle underlying transistors.

Unlike vacuum tubes, **transistors** are compact and require no power for heating the cathode. They are used in space exploration, computers, and the multitude of small appliances we use every day.

Integrated Circuits

In place of the tedious and intricate wiring once required for electronic equipment, **integrated circuits** can now be designed, photoreduced, and then reproduced in tiny wafers made of silicon from which the impurities have been removed. Several thousand circuits—transistors, resistors, connections—may be combined

Integrated circuit technology allows engineers to design extremely complex circuits, reduce them photographically, and reproduce them on silicon wafers (microchips) as little as $3/16$ inch square.

on a single silicon wafer ³/₁₆ inch square. These wafers, which are called microchips, may be inserted on circuit boards to become part of a larger circuit. In a computer, the microprocessor is a single chip containing all the circuits required to perform all the logic and arithmetic functions of the computer through a series of on/off switching.

Electronic Devices

Radio signals are carried by a kind of electromagnetic wave that oscillates at frequencies of around 1 million cycles per second. You would not be able to hear them, since these carrier waves oscillate too rapidly to vibrate the diaphragm by which a telephone receiver or a loudspeaker translates electrical waves into sound energy. Therefore, radio transmissions must be *modulated* before being broadcast. This means that the amplitude of the carrier waves is changed to the tempo of the sound waves they are to carry. Once it has been partially rectified as explained in the description of vacuum tubes, the modulated wave is able to operate a loudspeaker to reproduce the orig-

inal sound vibrations. This process of **amplitude modulation** is the basis of AM radio transmission (Figure 16.4).

In the **frequency modulation** system, the carrier wave has a constant amplitude, but its frequency is changed according to the sound waves. This system almost completely eliminates "static" and fading, which is one reason for the popularity of FM radio.

Figure 16.5

FREQUENCY MODULATION

Cathode Ray Tubes

The TV picture tube is one form of **cathode ray tube (CRT):** a glass container with a vacuum inside having a cathode and an electron gun that projects electrons at a luminescent screen. CRTs are also used in computer monitors, radar screens, and many kinds of monitoring instruments used in medicine and scientific research.

Basically, the picture tube on your television set is a refinement of the apparatus Thomson used in experimenting with cathode rays. Within the tube, electrons from a hot filament are accelerated electrically, and the narrow beam is made to

Figure 16.4

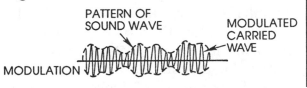

PATTERN OF SOUND WAVE

MODULATED CARRIED WAVE

MODULATION

In AM radio transmission, the amplitude of the carrier waves is modulated to match the amplitude of the sound waves being transmitted; in FM transmission, the frequency of the carrier waves is modulated to match the frequency of the sound waves.

A cathode ray tube (CRT) is an evacuated glass container in which an electron gun projects electrons at a luminescent screen, producing an image by means of a moving, bright spot of light.

pass between two pairs of plates (Figure 16.6). Voltages applied to one pair can deflect the beam vertically; those applied to the other pair can deflect it horizontally. Because the electrons have small inertia and high speed, they respond so fast that the path of the beam is marked by the luminous curve it traces on the fluorescent coating on the inside of the tube at the end you see.

Figure 16.6

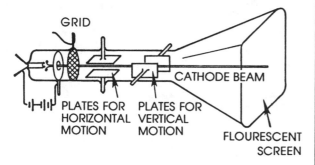

One pair of plates makes the beam move rapidly *across* the screen at regular intervals. The other jerks it *down* a short distance after each sweep. A spot of light moves on the screen the way your eyes move when you read a book. In about $1/30$ second, a rectangle of light is "painted" on the end of the tube. The incoming signal is applied to the grid in such a way as to vary its brightness, and the differing degrees of brightness reproduce the details of the picture.

The dots created on the screen by the incoming signal are called pixels, and the sharpness of a TV picture is dictated largely by the number of pixels in each vertical and horizontal row: The more pixels, the sharper the picture, just as a printed photograph is clearer the more dots used to form the image. Different countries have different standards for the number of dots used in commercial TV transmission.

At the TV station, the camera tube, using photocells, scans the scene and converts it into a succession of electrical impulses. These are amplified and transmitted on a carrier wave, as is the accompanying sound. The process is too complex to describe fully here. The electrical impulses that compose the TV signal can also be stored on magnetic tape—the entire VCR industry has arisen in just the few years since this technology was made practical for home use.

Another variation on the use of cathode rays is the **electron microscope,** a device for focusing cathode rays at a different point from where they originated. An electron lens does for cathode rays what a glass lens does for light rays (Chapter 11). In the electron microscope, the form and structure of objects placed in the path of the rays can be seen on a CRT. Magnifications of 10 to 100 times those possible with the best light microscopes can be obtained, revealing the structure of viruses, crystals, and even the DNA of cells.

With the scanning electron microscope, the surface structure of thicker specimens can be observed. Electrons are fired at the specimen and reflected back so that an image of the surface can be formed for viewing on a screen or storage in a computer.

Testing Your Knowledge

16.1 The fact that cathode rays can be swerved aside by both electric and magnetic fields shows that they
a. are small magnets.
b. carry an electric charge.
c. are electromagnetic waves.
d. contain atoms.

16.2 A metal plate located near a hot filament will acquire a negative charge because
a. the plate gives positive atoms.
b. the filament gives off protons.
c. the plate loses electrons by thermionic emission.
d. the filament releases electrons.

16.3 X rays are produced when electrons are
a. suddenly stopped.
b. sent through a wire.
c. neutralized.
d. made to leave a hot wire.

16.4 The action of a photocell is, in a certain sense, just the opposite of that of an X-ray tube. Explain.

16.5 How many electrons are released each second in a photocell when the current amounts to one-millionth of an ampere? (It takes 6.3 billion billion electrons to carry one coulomb of charge.)

GLOSSARY

absolute zero
> Temperature at which gases cease to exert pressure; zero degrees kelvin

acceleration
> The rate of change of velocity; that is, the change in velocity divided by the time it takes for the change to occur

acceleration due to gravity
> The constant rate of acceleration of falling bodies, ignoring air resistance or friction; 32 ft/sec or 980 cm/sec

acoustic interference
> Effect produced by crossing of sound waves produced simultaneously by two or more sources

alternating current (AC)
> The type of current that naturally results from the continued turning of a coil (electromagnet) in a fixed magnetic field

ampere

Rate of flow equal to 1 coulomb of electric charge per second or 6.3 billion billion electrons per second

ammeter

Instrument for metering the flow of electricity

amplitude

Distance between the crests and hollows of a sound wave; the greater the amplitude, the greater the intensity (loudness) of the sound

barometer

Instrument used to measure air pressure

boiling

Point in the heating of a liquid when the pressure of the vapor equals the pressure of the surrounding air, causing bubbles of vapor to form all through the liquid; 100 degrees C, 212 degrees F, 373 kelvin

British thermal unit (Btu)

Amount of heat (thermal energy) needed to raise the temperature of 1 pound of water by 1 degree F

buoyancy

Tendency of an object to rise (float) when submerged in a liquid

calorie

Basic unit of thermal energy in the metric system; kilocalorie (kcal) is used in dealing with large quantities of heat

capacitance

Rating, stated in ohms, of the ability of a conductor to take on more charge as its potential is raised

cathode ray tube (CRT)

Glass container in which a vacuum has been induced and an electron gun produces images by means of electrons fired at a luminescent screen; early researchers called electrons cathode rays

center of gravity

Single force equal to the entire weight of an object and considered to be acting at a given place

centimeter

Metric length measurement equal to 0.01 meter

Celsius system
System of temperature measurement in which the ice point is designated as 0 degrees and the steam point as 100 degrees; formerly called centrigrade system

centrifugal force
Force outward (away from the center) in a rotating object

centripetal force
Force inward (toward the center) tending to keep an object on a circular path

coefficient of heat conduction
Number expressing the proportional ability of a material to conduct heat, silver being the standard with a conduction coefficient of 100

coefficient of linear expansion
Increment by which substances expand when heated from ice point to steam point; different for each substance

colloidal suspension
A stable mixture of a solid and a liquid, for example, ink

conduction
The handing-on of molecular motion through a substance; one way that heat passes from one material to another

conductor
A material that allows electric charge to pass freely; that is, with little resistance

convection
The mass movement of a heated liquid or gas

coulomb
Unit of the quantity of electrical charge

decibel (dB)
Unit for measuring the intensity of sound

degree
An increment of temperature

density
The mass of a substance divided by its volume

diffraction
The bending of waves when they pass near the edge of an obstacle or through a small opening

diffuse reflection
Reflection from a rough or irregular surface

direct current (DC)
>   Type of current in which the charge flows continually in one direction; "smoother" than AC

displacement
>   Weight or volume of a fluid that has been replaced by a floating body

Doppler effect
>   Tendency of sound to increase in pitch as listener and sound approach, then fall off sharply as they move apart again

dry cell
>   Type of voltaic cell used in flashlights, toys, etc.

electric current
>   An electric charge in motion

electric field
>   Space in the vicinity of charged objects in which their force or some part of it is exerted

electrical circuit
>   A source of potential difference combined with a series of conductors

electrical resistance
>   Opposition of a conductor to electric current

electromagnet
>   Device for increasing electromagnetic force by means of a soft-iron core

electromagnetic induction
>   Production of electric current by means of coil and magnet

electrostatic induction
>   Inducing electric charge in a neutral body by bringing it near a charged one

element
>   A chemical that cannot be further broken down into component chemicals

emulsion
>   Elobules of one liquid suspended in another, for example, cream (fat in water)

energy
>   In classical physics, activity that can bring about changes in matter, for example, light, heat, and electricity

equilibrium
> Point when the resultant force acting on an object is zero; rest or balance

erg
> Metric unit expressing work in centimeters—squared grams per second squared (cm²g/sec²)

evaporation
> Escape of molecules from the surface of a liquid

Fahrenheit scale
> System of temperature measurement in which the ice point is 32 degrees and the steam point is 212 degrees

fiber optics
> Light transmitted through fibers of glass or plastic using refraction to "bend" it around corners; used in communication and in medical diagnosis

flotation
> The relative capacity of a material or an object to float

foot-candle
> Unit used to express illumination or radiant energy

force
> Push or pull exerted on or by an object

force vector
> The amount or magnitude of a force plus its direction

frequency
> Number of vibrations per second of a wave form of energy

gas
> A state of matter that has neither definite volume nor definite shape

gravitational potential energy (GPE)
> Energy of position, that is, the amount of work an object can do by returning to its original position—for example, a coiled spring or a raised weight

grounding
> The loss of charge that occurs when a charged object is connected to a very large body

heat-work equivalent
> The rate of exchange between mechanical energy and thermal energy; whenever a given amount of mechanical energy disappears, a fixed quantity of thermal (heat) energy appears in its place

heat of fusion

Quantity of thermal energy required to melt a gram of a given substance without producing any change in temperature; different for each material

heat of vaporization

Quantity of thermal energy (heat) carried away from a liquid for each gram of liquid that it vaporizes; different for each material

hertz (Hz)

Unit referring to wave cycles per second; 1 Hz equals 1 cycle per second

ice point

In the Celsius system, the level of a mercury thread in a thermometer placed in a mixture of ice and water; 0 degrees C

illumination

The radiant energy (light) falling on any given unit area

index of refraction

The ratio of light's speed in a vacuum to its speed in a given material

infrared

Light waves falling beyond the red end of the visible spectrum

insulator

A nonconductor; material that impedes passage of electric charge

integrated circuit

Miniaturized circuit etched photographically on semiconductor "Chip"

joule

Metric unit of work equal to 10 million ergs

kelvin

Temperature increment of the Kelvin, or absolute, scale of temperature; formerly called degree Kelvin

Kelvin scale

Also, absolute scale. Temperature scale using absolute zero (-273 degrees C) as its starting point

kilometer

1000 meters; metric length measurement

kinetic energy (KE)

Energy of a moving body

lens

Device, often plastic or glass, that changes the curvature of light waves, usually in order to form an image

liquid

A state of matter in which the matter has no shape of its own but takes the shape of whatever it is in

longitudinal wave

Waves that move disturbed particles along the line in which the waves are moving

luminous intensity

The strength of a light source measured by the standard candle

machine

Any device, whether or not motorized, by which energy can be transferred from one place to another or one form to another

magnetic field

The space around a magnet in which its effects are exerted

magnetic flux

The total number of lines of force passing through any circuit located in a magnetic field

magnetic induction

Temporary transferring of magnetism from a permanent magnet to another material

magnetic poles

The two ends of a magnet, north and south

magnetism

Ability to attract iron and certain other metals with a similar molecular structure

mass

Amount of matter an object or substance contains

matter

In classical physics, anything that takes up space

mechanical energy

The amount of work an object (body) can do

metric system

System of measurements in which all fundamental units are multiples of 10

momentum

The propulsive force of a body

oscillate

Move back and forth about a center; vibrate

penumbra

The outer, somewhat lighted, part of a large shadow

photoelectric effect

Freeing of electrons from a solid substance by shining light on it

physics

The study of matter and energy

polarization of light

Passing a beam of light through a material that holds back all the wavelengths except those lined up with the material's own grain

potential difference

Difference in electrical charge between two objects; a charge will tend to move from the area of higher potential to the area of lower potential

power

The rate of doing work

pressure

The amount of force exerted by an object on the area of the surface on which it acts

primary colors

In the spectrum, the three colors from which every other color of light can be mixed: red, green, and blue-violet

quantum

The energy unit that makes up light

radiation

Movement of wave forces out from a center; also, the process in which heat is transferred by waves traveling through space

refraction

Change in the direction of rays in going through a boundary when coming in at an angle to the normal

relay

Electromagnetic device that allows a weak current to open and close a circuit in which a stronger current flows

resistance

The capacity of an object or material to impede motion

resonance

Process by which sound vibrations build up

resultant

The single force of definite magnitude and direction that is the combined effect of all forces acting on a body

second

Basic unit of time in both the English and the metric systems

semiconductor

Solid-state device having medium resistivity used to transmit and amplify electronic signals

solid

A state of matter that has a definite shape and volume

specific heat

The fractional number assigned to a substance to designate its heat energy, for which the standard is water, to which a number of 1 is assigned

spectrum

The sequence of colors that results when light is dispersed through a prism

speed

The rate at which something moves

standard candle

Unit of measurement of the strength of a light source

static electricity

An electric charge resting on an object

steam point

The position of a mercury thread in any given thermometer placed in the steam rising from boiling water; 100 degrees C, 212 degrees F, 373 kelvins

temperature

Degree of hotness or coldness of an object or an environment

thermal energy

Heat energy

time

A continuum along which events move from the past through the present and into the future

torque

Ability of a force to produce rotation

transformer

A device that changes voltage and current values

transverse waves
Waves that move disturbed particles perpendicularly to their line of advance

ultrasonic frequencies
Frequencies above the hearing range of humans, especially those of several hundred thousand Hz

ultraviolet
Light waves falling beyond the violet end of the visible spectrum

umbra
The central, totally lightless part of the shadow cast by a very large source of light

vacuum
A space with no matter in it

velocity
Speed plus direction

volt
A measure of potential difference equal to 1 joule per coulomb

voltaic cell
Unit of a wet-cell storage battery

voltmeter
Instrument for metering electricity by passing current through a rectangular coil having a high resistance

volume
Amount of three-dimensional space occupied by matter

watt
A working rate of 1 joule per second

weight
The pull of earth's gravity upon an object

work
Transfer of energy to an object by the application of force

X rays
Waves having shorter length than the ultraviolet, emitted by radioactive substances

APPENDIX

Appendix A: Important Formulas and Relations

Measurement

Density of a substance: $D = \dfrac{M}{V}$, where D is the density, M is the mass of a sample of the material, and V is the volume of that sample.

Force

Torque, or turning effect, of a force about a given pivot point: $T = Fh$, where T is the torque, F is the amount of the force, and h is the perpendicular distance from the pivot to the line of the force.

Equilibrium of torques (condition for no rotation): Sum of torques tending to turn body in one direction = Sum of torques tending to turn it in the opposite direction.

Law of Gravitation: $F = \dfrac{Gm_1m_2}{d^2}$, where F is the force of attraction, m_1 and m_2 are the two masses, and d is their distance apart.

Motion

Average speed of motion: $v = \dfrac{d}{t}$, where v is the average speed, d is the distance covered, and t is the elapsed time.

Acceleration: $a = \dfrac{v}{t}$, where v is the change in speed and t is the time required to produce that change.

Newton's second law: $\dfrac{F}{W} = \dfrac{a}{g}$. Here F is the force acting on a body of weight W, g is the acceleration due to gravity, and a is the acceleration of the body's motion. F and W must be measured in the same units.

Momentum: $M = mv$, where M is the momentum, m is the mass of the body, and v is its velocity.

Work and Energy

Work done by a force: $W = Fd$. W is the amount of work done, F is the magnitude of the force, and d is the distance moved in the direction of the force.

Kinetic energy of a moving body:

$$\text{KE}_{\text{ft-lb}} = \frac{m_{\text{lb}} v^2_{\text{ft}^2/\text{sec}^2}}{64}$$

$$or \ \text{KE}_{\text{gm-cm}} = \frac{m_{\text{gm}} v^2_{\text{cm}^2/\text{sec}^2}}{2}$$

where KE is the kinetic energy, m is the mass, and v is the speed of the body.

Power: $P = \dfrac{W}{t}$, where P is the average power expended, W is the amount of work done, and t is the time required to do it. In horsepower,

$$P_{\text{HP}} = \frac{W_{\text{ft-lb}}}{550 \times t_{\text{sec}}} .$$

Heat

Celsius and Fahrenheit temperatures: Readings on the two scales are related by $F = \dfrac{9}{5}C + 32$, where C is any temperature on the Celsius scale and F is the corresponding one on the Fahrenheit scale.

Thermal Energy

Quantity of heat: $Q = smt$. Here Q is the quantity of heat taken on or given off, s is the specific heat of the material, m is the mass of the body, and t is its temperature change.

Heat–work equivalent: 1cal is equivalent to 4.18 joules, or 1 Btu is equivalent to 778 foot-pounds.

Pressure

Boyle's law: If the temperature of a gas remains constant, $\dfrac{V_1}{V_2} = \dfrac{p_2}{p_1}$, where p_1 and V_1 are the pressure and volume, respectively, in one case, and p_2 and V_2 are the values in another.

Pressure: $p = \dfrac{F}{A}$, where p is the pressure acting on a surface, F is the total force and A is the area to which it is applied.

Pressure beneath the surface of a liquid: $p = hD$, where p is the pressure at any point, h is the depth of that point below the surface, and D is the density of the liquid.

Archimedes' law: Buoyant force on a body immersed in a liquid = Weight of liquid displaced by the body.

Wave Motion

Wave equation: $V = nl$, where V is the speed of the wave, n is its frequency, and l is its wave length.

Light

Illumination produced by a small light source on a surface held perpendicular to the rays:

$E = \dfrac{C}{d^2}$ E is the illumination, C is the intensity of the source, and d is its distance from the illuminated surface.

Index of refraction: $n = \dfrac{c}{V}$, where n is the index of refraction of a material in which the speed of light is V, and c is the speed of light in a vacuum.

Location of image formed by a converging lens:

$\dfrac{1}{p} + \dfrac{1}{q} = \dfrac{1}{f}$, where p is the distance of the object from the lens, q is the distance of the image from the lens, and f is the focal length of the lens.

Size of the image: $\dfrac{h_1}{h_o} = \dfrac{q}{p}$, where h_1 is the height of the image, h_o the height of the object, q is the image distance, and p is the object distance.

Electricity

Strength of an electric current: $I = \dfrac{Q}{t}$, where I is the current strength, Q is the total quantity of charge passing any point in the conductor, and t is the time during which it passes.

Ohm's law: $I = \dfrac{V}{R}$, where I is the strength of the current flowing in a conductor, V is the PD applied to its ends, and R is its resistance.

Resistors in series: $R = R_1 + R_2 + R_3$, etc.

Resistors in parallel: $\dfrac{1}{R} = \dfrac{1}{R_1} + \dfrac{1}{R_2} + \dfrac{1}{R_3}$ etc. Here R is the combined resistance, and R_1, R_2, etc. are the separate values.

Power expended in an electric appliance:

$$P_{\text{watts}} = I_{\text{amp}} \times V_{\text{volts}}.$$

Heat produced in a conductor by a current:

$$Q = 0.24\ I^2 R t.$$

Here Q is the quantity of heat, in calories, I the current in amperes, R the resistance of the conductor, in ohms, and t the time the current flows, in seconds.

Transformer: $\dfrac{V_S}{V_P} = \dfrac{n_S}{n_P}$, where V_P is the voltage in the primary coil, V_s that in the secondary, and n_P and n_s are, respectively, the numbers of turns in each.

Relativity

Mass-energy equivalence: $E = mc^2$, where E is the energy, in ergs, equivalent to a mass m, in grams, and c is the speed of light in centimeters per second.

Appendix B: Major Principles and Laws

Law of Gravitation: Any two kinds of bodies in the universe attract each other with a force that is directly proportional to their masses and inversely proportional to the square of their distance apart:

$$F = \frac{Gm_1m_2}{d^2}$$

(*Chapter 3*)

Law of Acceleration: The amount of acceleration is directly proportional to the acting force and inversely proportional to mass: $F = ma$
(*Chapter 4*)

Law of Constant Acceleration (Newton's second law): A body acted upon by a constant force will move with constant acceleration in the direction of that force:

$$\frac{F}{W} = \frac{a}{g}$$

(*Chapter 4*)

Law of Inertia (Newton's first law): Every body remains in a state of rest or uniform motion unless acted upon by forces from the outside.
(*Chapter 4*)

Law of Conservation of Momentum (Newton's third law): When an object is given a certain momentum in a given direction, another body or bodies will acquire an equal momentum in the opposite direction.
(*Chapter 4*)

Law of Conservation of Mechanical Energy: Ideally, in the absense of friction, energy stored in a machine remains constant and work done by the machine is equal to the work done on it.
(*Chapter 5*)

Law of Conservation of Energy: Energy cannot be created or destroyed; it changes forms but does not cease to exist.
(*Chapter 7*)

Bernoulli's Law: A moving stream of gas or liquid exerts less sidewise pressure than if it were at rest.
(*Chapter 7*)

Archimedes' Law of Buoyancy: Any object immersed in a liquid appears to lose an amount of weight equal to that of the liquid it displaces; therefore, buoyant force on a body immersed in a liquid equals the weight of liquid displaced.
(*Chapter 8*)

Boyle's Law: The volume of a gas is inversely proportional to the pressure (when t is constant):

$$\frac{V_1}{V_2} = \frac{P_2}{P_1}$$

(*Chapter 8*)

Law of Reflection: The angle of incidence equals the angle of reflection.
(*Chapter 10*)

Lenz's Law: The direction of an induced current is always such that its magnetic field opposes the operation that causes it.
(*Chapter 15*)

Ohm's Law: The current in a wire is proportional to the potential difference between the ends of the wire:

$$I = \frac{V}{R}$$

(*Chapter 15*)

Appendix C:
Answers and Solutions to
Testing Your Knowledge

Chapter 2

2.1 Since there are 36 in to a yard, 38.7 yd will contain 38.7 × 36 = 1390 in (rounded off to three significant digits).

2.2 There are 39.4 in to one meter, so 1.34 m will amount to 1.34 × 39.4 = 52.8 in

2.4 Since 2.54 cm = 1 in, this will be 25.4 threads per inch.

2.5 3000 m = 3000 × 39.4/12 ft. The cost per foot is 0.14 cent, so the above length will come to 3000 × 39.4 × 0.14/12 = 1379 cents or $13.79.

Chapter 3

3.1 c; force is a vector.

3.2 a.

3.3 c.

3.4 d, for then the amount of the resultant force is the sum of the two.

3.5 b.

3.6 Compare the height of the center of gravity in the two cases.

3.7 If the load is 4 ft from the left-hand end, then taking torques around this end give us 150 × 4 = R × 9, or R = 66.7 lb, where R is the force with which the right-hand end is supported. Then the force at the other end must be simply 150 − 66.7 = 83.3 lb.

3.8 2.7 lb.

3.9 Since the force is inversely proportional to the square of the distance, it would be reduced to $\frac{1}{3}^2 = \frac{1}{9}$ of its present amount.

3.10 Substituting in the gravitational formula,

$$\frac{(0.000000000033) \times (15000 \times 2000)^2}{(150)^2}$$

or F = 1.3 lb

Chapter 4

4.1 The first part of the trip takes 1/12 hour. The average speed is the total distance divided by the total time, or

$$\frac{3\frac{1}{2}}{\dfrac{1}{12} + \dfrac{1}{4} + \dfrac{1}{10}} = \frac{105}{13} = 8.1 \text{ mi/hr.}$$

4.2 The acceleration on the moon will amount to 32/6 = $5\frac{1}{3}$ ft/sec². At the end of 2 sec, the stone will have gained a speed of 2 times this figure, or 10.7 ft/sec.

4.3 During the first second, the average speed is ½ (0 + 32) = 16 ft/sec, so the body goes 16 ft during this time. The speed at the beginning of the second second will be 32 ft/sec, and at the end of the second second, it will be 64 ft/sec. Hence the average speed in this interval will be ½ (32 + 64) = 48 ft/sec. Therefore, the body will go 48 ft during the second second, or three times as far as in the first second.

4.4 Average speed would be increased, time required decreased.

4.5 Since the bullet "drops off" as it goes along, the aim must be high.

4.6 If the accelerations involved are high, what about the forces? See Newton's second law.

4.7 The hammer is brought to rest in a very short time interval. What is the magnitude of its acceleration during this time? Consider Newton's second law.

4.8 The gun has a much greater mass than the bullet; hence what must be true of its recoil speed? Think about Newton's third law, the law of conservation of momentum.

4.9 On ice, the friction is very small, so there is no resisting force against the horizontal movement of our foot.

4.10 The ferry has a much larger mass. What does this tell us about its momentum?

Chapter 5

5.1 b.

5.2 a.

5.3 b.

5.4 40 pounds.

5.5 80 pounds.

5.6 At each stage, work done *on* machine equals work done *by* machine. Thus the two rates of working are equal.

Chapter 6

6.1 Celsius is larger. The ratio is 5 Celsius degrees for every 9 Fahrenheit degrees. There are $212 - 32 = 180$ Fahrenheit degrees between the ice and steam points, while there are 100 Celsius degrees in the same interval.

6.2 From the relation on p. 48 we find $C = 37.0°$. The absolute value is $37 + 273 = 310° K$.

6.3 The coefficient of expansion is less than half that of ordinary glass, therefore it will shrink half as much for the same temperature change. It will be subjected to half as much stress and be less likely to break.

6.4 The engine parts expand as the engine warms up. As they expand they make noise until the temperature stabilizes.

6.5 Larger. The metal expands outward from the center at every point. Another way to look at it: What must happen to any imaginary band of metal surrounding the cavity as the metal is heated? Therefore, what happens to the size of the cavity?

6.6 The temperature of the piston rises by 160 Celsius degrees. By Table 4, the fractional increase in length for aluminum, per degree, is 0.000024, so the actual increase in length will be $0.000024 \times 160 \times 2\frac{3}{4} = 0.011$ in, about.

6.7 b.

6.8 c.

6.9 b.

6.10 d.

6.11 a.

Chapter 7

7.1 Since they have very small mass,

the quantity of heat they carry is small in spite of their high temperature.

7.2 Using $Q = smt$ we get $Q = 0.11 \times 5 \times 265 \times 146$ Btu.

7.3 Heat needed to
melt the ice: 144 Btu
Heat to raise the resulting water by 180°F to its boiling point: 180
Heat to change the water to steam: 970
                              Total: $\overline{1,294}$ Btu

7.4 (a) Any heat supplied after the water has begun to boil is carried off in the steam produced instead of going toward a further increase of temperature. (b) Heat is needed to melt ice; thus heat must be given up when water freezes.

7.5 Although the steam and water each have the same temperature, the steam contains much more heat energy (calories).

7.6 No. If the air is at the same temperature as the object, the only way cooling could be produced would be by evaporation.

7.7 Condensed from vapor in the air.

7.8 The ice formed eventually evaporates. Melting is not involved.

7.9 Do the brakes become warmed?

7.10 Since 778 ft lb is equivalent to 1 Btu, the quantity of heat amounts to $160 \times 3900/778 = 802$ Btu.

7.11 Calling the distance in feet d, the work needed will be $3000d$ ft lb. One fourth of 30,000 Btu is equivalent to $30,000 \times 778/4$ ft lb. Setting the two equal and solving, $d = 1,945$ ft.

7.12 Higher.

7.13 How is the heat taken from the inside of the box disposed of? Also, what is the effect of the driving motor of the gas flame?

Chapter 8

8.1 Since pressure increases with depth, the pressure would be greater under the hump and as a result the water would flow outward from it in all directions, until everything is at the same level.

8.2 This, too, is due to the fact that pressure increases with depth, so the tank walls must be made progressively stronger toward the bottom.

8.3 See answer to preceding question.

8.4 No. The pressure is the same in each case, since the depth is the same. The *extent* of the body of water makes no difference.

8.5 Use the relation $p = hD$: $p = 30 \times 1 = 30$ gm/cm².

8.6 The pressure is given by $p = 100 \times 64 = 6,400$ lb/ft². The total force is found by multiplying by the area: $F = 6400 \times 1500 = 9,600,000$ lb, 4,800 tons force.

8.7 c, since 100/80 = 1.25.

8.8 c, since water is denser than gasoline.

8.9 c.

8.10 c.

8.11 According to Table 3, aluminum and lead will, while gold will not.

8.12 Cork is much less dense than water—about ¼ as much.

8.13 Yes it will change. The ship will ride higher because salt water is denser than fresh, and the hull will have to displace less salt water.

8.14 The boat will have to displace an additional 20 tons of water, whose volume, given by $V = M/D$ is $20 \times 2000/62.4 = 641$ ft³. With a 5,000 ft² area, the thickness of this layer of water would have to be $641/5000 = 0.13$ ft, or about an inch and a half.

8.15 From Table 2.3 the ratio of the density of ice to that of sea water is $57/64 = 0.89$; therefore, about 89 percent of the bulk of an iceberg is under water.

8.16 The computation goes: $p = 30 \times 850/1728 = 14.8$ lb/in².

8.17 The difference in pressure is $14.7 - 5.0 = 9.7$ lb/in² and the area of the lid is $\pi(2.5)^2 = 19.6$ in². The whole force is then $19.6 \times 9.7 = 190$ lb.

8.18 By Boyle's law,
$$\frac{100}{7.35} = \frac{p_2}{14.7}, \text{ so } p_2 = 200 \text{ lb/in}^2.$$

8.19 Archimedes' principle says that the buoyant force is equal to the weight of the displaced air, which is $4000 \times 0.08 = 320$ lb. Also, the hydrogen weighs $4000 \times 0.08 = 320$ lb. Also, the hydrogen weighs $4000 \times 0.0055 = 22$ lb. This, together with the bag, makes a total weight of 72 lb. The difference, $320 - 94 = 226$ lb, is the "pay load."

8.20 Strictly speaking, you do not *suck* the air in, you merely enlarge your lungs and normal outside air pressure *pushes* more air into them.

8.21 The suction cup seals around the edge. As you pull, you increase the volume and decrease the pressure inside the cup, and the force you feel is due to the difference between the atmospheric pressure and the pressure inside the cup.

8.22 Refer back to the remarks on p. 55.

8.23 Decide what effect the pumping will have on the resultant pressure on the balloon.

8.24 The parachute in effect greatly increases the cross-section area of the falling body. How does this affect the resistance to motion through the air?

8.25 Between the boats there is what amounts to a swift current of water toward the stern. Recall Bernoulli's principle.

Chapter 9

9.1 The soldiers at the end hear the music later than the soldiers at the front because it takes time for the sound to travel from the front of the line to the back.

9.2 Is a tune played by a band recognizable by a listener even if he is some distance away? What can you conclude from this?

9.3 $d = \frac{1}{2}(3.5)(4700) = 8,200$ ft.

9.4 Dividing the distance in feet by 1100 and by 86,400 (the number of seconds in a day) gives about 13 days as the result.

9.5 b.

9.6 The speed in ft/sec is $5280/4.8 = 1100$. According to Table 8, the speed in air at 20° C is 1126 ft/sec. The speed of sound decreases about

2 ft/sec for each degree drop in temperature, so the temperature must have been 13° lower than 20°, or 7° C.

9.7 1100/256 = 4.3 ft.

9.8 The frequency stays the same, even if the waves pass into another material. Since $V = nl$, going into water where V is about 4 times as great will make l about 4 times as great.

9.9 a.

9.10 b.

9.11 c.

9.12 b.

9.13 c.

9.14 An echo is caused by hearing a sound again after it "bounces" off another surface. Reverberation is a sound from an object being made to vibrate by the waves from another sound.

9.15 One half and twice this value, respectively.

9.16 To increase their weight.

9.17 What happens to the length of the air column in the jar? Think about the pitch of organ pipes of different lengths.

9.18 These are doubtful cases, but the tendency is to call the piano a stringed instrument and to refrain from putting the voice in any of these classes.

Chapter 10

10.1 c. It is the only self-luminous source.

10.2 The number of minutes is given by 93,000,000/(186,000 × 60) = 8⅓.

10.3 The image will become 12/8 = 1.5 times as large when he comes to the position 8 ft from the camera.

10.4 Illumination being inversely as the square of the distance, it will amount to $1/(⅓)^2 = 9$ times as much.

10.5 c.

10.6 c.

10.7 a. Make a diagram showing a side view of the situation. A ray coming from the man's toe to the mirror and then to his eye must hit the mirror in such a way that the angles of incidence and reflection are equal. The same for a ray from the top of his head and to his eye. The mirror will have to extend from one of these places on the wall to the other. How big is this distance in terms of the man's height? Does his distance from the wall make any difference in the result?

10.8 It has no dimensions, since it is the ratio of two speeds. It is merely a pure number.

10.9 Consider the type, size and positions of the images that can be formed.

10.10 Bifocal eyeglasses have lenses that are able to correct the vision of a person who requires two different focal lengths, one for close vision and the other for more distant vision.

10.11 $1/p + 1/10/5 = 1/10$, or $1/p = 1/210$, $p = 210$ in, which is 17½ ft.

10.12 2500/14 = 179 in, or nearly 15 ft.

10.13 The area of a .2-in circle is .0314 sq in. The area of a 200-in circle is 31,400 sq in. or 1,000,000 times as much.

Chapter 11

11.1 b. Consider the directions in which the various rays come to the eye.

11.2 a. See Fig. 11.2.

11.3 c. Only white light will give the true colors.

11.4 d. The first observation shows that it reflects red light, the second shows that it does not reflect blue at all. Hence it cannot be either blue or white.

11.5 a.

11.6 $c = 30,000,000,000$ cm/sec, about. Then $n = c/l = 500,000,000,000,000$ (500 trillion) vibrations per second.

11.7 (a) Moonlight is merely reflected sunlight. (b) Line spectrum. (c) Continuous spectrum.

11.8 The smallest visible wavelength is purple, therefore the lines of the spectrum would be displaced toward the violet end of the spectrum.

11.9 The wires running each way act as a coarse diffraction grating.

11.10 Are these colors due to pigments or to something else?

Chapter 12

12.1 What kind of charge will be induced on the near end of the object? The force at the near end will dominate because of the smaller distance.

12.2 After touching, the force is no longer due to induced charges only. What else happens?

12.3 Is there, in a sense, any rubbing involved?

12.4 The strap acts as a ground, causing the charge to be conducted into the ground.

12.5 (a) Note that they have opposite kinds of charge. (b) The act of touching leaves a balance of 1 billion electron charges, and these are shared equally by the two, leaving half a billion (500 million) on each. The force is one of repulsion.

Chapter 13

13.1 Dividing the charge by the time gives $1/0.0002 = 5000$ amp.

13.2 No; two *different* metals must be used.

13.3 Nine storage cells in series would have a total PD of about 18 volts. Therefore $18/1.5 = 12$ dry cells would be needed.

13.4 A fully charged one.

13.5 The storage battery actually stores chemical energy that is put into action when the circuit is completed.

13.6 The resistance is proportional to length divided by cross-section. The cross-section, in turn, is proportional to (diameter)^2. Then, if length and diameter are both doubled, the resistance will be multiplied by $2/2^2 = \frac{1}{2}$; it will be half as much as before.

13.7 The PD across each lamp is $120/8 = 15$ volts. Then, using $R = V/I$, R turns out to be 75 ohms.

13.8 Call the value of the resistor r. Then Ohm's law for the resistor and appliance together is $2 = 120/(25 + r)$ and $r = 35$ ohms.

13.9 The current flowing in the smaller resistance will be 3 times that in the larger; that is, ¾ of the *total* current goes through the former.

13.10 The equivalent resistance of the two coils in parallel is given by $1/R = ⅓ + ⅙$, or $R = 2$ ohms. Then, using Ohm's law for the whole circuit, $I = 12/(2 + 2) = 3$ amp, and this is also the current in the 2-ohm coil.

13.11 The current in the 3-ohm coil will be twice as great as that in the 6-ohm coil, and since the total current is 3 amp, ⅔ of this, or 2 amp will pass through the 3-ohm coil.

13.12 If we substitute the value of I from Ohm's law ($I = V/R$) into the expression for power ($P = IV$), we get $P = V^2/R$. Putting in the numbers, $30 = 144/R$, or $R = 4.8$ ohms.

Chapter 14

14.1 c. 14.4 c.
14.2 b. 14.5 c. (see Fig. 14.4)
14.3 b.
14.6 c.
14.7 b. Since 1 watt is 1 joule per second, the number of joules of energy expended in 5 minutes is $100 × 5 × 60 = 30,000$. The quantity of heat delivered when the temperature rises t C° is $Q = 1 × 225 × t$ cal, or $225 × 4.18 × t$ joules. Setting this equal to 30,000 and solving for t, we get 32C °.

14.8 b.
14.9 a.
14.10 d.
14.11 Attract each other when currents are in same direction; repel when in opposite directions.

14.12 Since the coil has 9 times as much resistance as the shunt, 1/10 of the total current, or 1 amp, will flow in the coil.

14.13 The coil constitutes 0.1/500.1, or about 1/5000, of the total resistance, and so the PD across the coil will be $10/5000 = 0.002$ volt.

14.14 None, since *both* the current in the field magnets and that in the coils will be reversed. Convince yourself by sketching the field lines.

Chapter 15

15.1 Remembering that the field is uniform, is there any *change* in the flux through the hoop when moved as described?

15.2 North of the equator the lines of the earth's field have a downward direction. In order to oppose the motion of the wire (Lenz's law), the lines of force of the induced current would also have to be downward on the front side of the wire, so the current would have to be toward the west (left-hand wire rule). South of the magnetic equator, the result would be just the opposite.

15.3 It is increased in the same proportion.

15.4 5 to 1.

15.5 According to the relation on p. 165, the secondary current will be 50 amp.

15.6 The electromagnetic forces act only when current actually flows in the windings, since these forces are really between two magnetic fields—that of the field coils and that of the rotating coils. The work done against this opposition accounts for the energy of the current produced.

15.7 With a back voltage of 45 volts, the actual voltage applied to the coils is 50 − 45 = 5 volts. Then, by Ohm's law, the current will amount to 2.5 amp. If the motor is not turning, the back voltage will be absent and the current in the coils will be 50/2 = 25 amp. This would likely burn out the windings, since the heating effect is proportional to I^2, and so would become 100 times as great as normal.

Chapter 16

16.1 b.

16.2 d.

16.3 a.

16.4 Both involve the interaction between radiation and electrons. Can you state explicitly how each operates in these terms?

16.5 One millionth amp is 6,300,000,000,000,000,000/1,000,000 = 6,300,000,000,000 (6.3 trillion) electrons per second.

NOTES:

INDEX

Page numbers followed by (t) indicate tables.

longitudinal wave,   85, 181
loudness,   93
luminous intensity,   101, 102, 181

M
machines,   45, 181
magnetic field,   150, 181
magnetic flux,   160, 181
magnetic induction,   150, 181
magnetic lines of force,   151
magnetic poles,   149, 151, 181
magnetism,   148–156, 181
  induction of,   150, 181
magnetite,   148
mass,   18, 181
matter,   10, 181
  characteristics of,   11
Maxwell, James Clerk,   104
Maxwell's theory of light,   104
Mayer, J. R.,   65
measurement,   13–21
  of light and color,   121
  of electric current,   137
  of electric power,   144
  of illumination,   102
  of weight and mass,   18
mechanics,   40
mechanical energy,   42, 181
meter,   15
metering electricity,   155
metric system,   15, 181
metric ton,   19
Michelson, Albert A.,   14
microchips,   171
microprocessor,   171
microscope, compound,   114
  electron,   172
milligram,   19
milliliter,   17
millimeter,   15
mirror principle,   106
mirrors,   107
modern physics,   10
momentum,   36, 181, 186
motion,   31–38, 185

music,   95
myopia,   114

N
nearsightedness,   114
negative charge,   131
Newton, Sir Isaac,   26, 34, 117
  corpuscular theory of,   103
  first law of motion of,   34, 35, 188
  law of conservation of momentum of,   34, 37, 188
  law of constant acceleration of,   35, 188
  law of gravitation of,   26, 188
  law of inertia of,   35, 188
  second law of motion of,   34, 35, 186, 188
  theory of light of,   117
  third law of,   34, 37, 188
noise,   92
nonconductors,   53, 132
nonreflecting glass coatings,   124
normal (of light),   105
north magnetic pole,   151

O
Oersted, H. C.,   152
ohm,   141
Ohm, G. S.,   140
Ohm's law,   140, 187, 188
optics, fiber,   106, 179
  wave,   117–126
oscillation,   85, 182

P
parallel circuit,   143
pendulum,   43
penumbra,   100, 182
permanence of matter,   11
photoelectric effect,   168, 182
photons,   104
physical changes of matter,   10
physics,   9, 182
pigment,   119
pinhole camera,   100
pitch,   91